The Mars Book

*A Guide to Your
Personal Energy and Motivation*

Donna Van Toen

First published in 1988 by Samuel Weiser, Inc.
Copyright 1988 by Donna Van Toen

Revised Edition Copyright 2012 by Donna Van Toen

No part of this book may be reproduced or transcribed in any form or by any means, electronic or mechanical, including photocopying or recording or by any information storage and retrieval system without written permission from the author and publisher, except in the case of brief quotations embodied in critical reviews and articles. Requests and inquiries may be mailed to: American Federation of Astrologers, Inc., 6535 S. Rural Road, Tempe, AZ 85283.

ISBN-10: 0-86690-588-X
ISBN-13: 978-0-86690-588-6

Cover Design: Jack Cipolla

Published by:
American Federation of Astrologers, Inc.
6535 S. Rural Road
Tempe, AZ 85283

Printed in the United States of America

Contents

Introduction	v
Introduction to the Revised Edition	vii
Chapter 1, Everyday Mars	1
Chapter 2, Mars and Your Love Life	27
Chapter 3, Occupational Motivation	43
Chapter 4, Energy from Your Environment	53
Chapter 5, Mars and Your Luck	67
Chapter 6, Special Cases	81
Chapter 7, Progressed Mars	93
Chapter 8, Transits of Mars	109
Chapter 9, The Mars Return Chart	123
Chapter 10, Mars and Your Inner Motivation	131
Chapter 11, Esoteric House Positions	137
Chapter 12, Esoteric Aspects	145
Chapter 13, Sample Mars Analysis	155
Afterword	165
Appendix I, Mars Affinities	167
Appendix II, Sun-Mars Quickie Synastries	169
Appendix III, Mars Positions for Various Cities	183

Introduction

The planet Mars has been neglected by astrologers and taken for granted by students for years. Hence this book. Mars is a crucial key to understanding what makes you tick. It shows where and how you use your energies, as well as the results of your energies. It plays a major role in what happens to you in life. For this reason, it demands attention. A good understanding of Mars is absolutely necessary if you're to understand who you are and where you're going in life.

Mars rules Aries and has a natural affinity with the first house in your chart. It's influential in the house it's found in or in any house having Aries on the cusp or containing an intercepted Aries. The symbol for Mars is the cross of matter over the circle of spirit, suggesting that "wants" of various types are the dominant theme where Mars is active, with emotions and ideas playing a role only insofar as they are channeled into action designed to help you get what you want.

Mars represents courage and energy. It indicates heat and fevers, scars, wounds, and accidents. It describes your temper. In a woman's chart, it shows her opinion of what men should be like and therefore influences the type she'll be attracted to. In a man's chart, it influences his concept of what masculinity should entail and therefore helps describe the sort of lover he is. The Mars influence enables us to take action. It encourages us to be energetic and forceful, or it can cause us to be timid and passive in our approach to life. In other words, Mars shows your ability to assert yourself as an individual. Its energy says, "I want to act in a specific way, along specific lines, to get specific results, and I'll fight any interference in whatever way I can."

A strong Mars says you're able to take care of yourself and aren't easily threatened by your environment or the people in it. A weak Mars, on the other hand, says you may have problems—excessive aggressiveness or excessive passivity, over-developed or underdeveloped self-preservation instincts, lack of awareness of your energy patterns, romantic, sexual, or health problems—that make you less comfortable with your environment and therefore less confident of your ability to get what you want.

To understand Mars, we need to look at the influences it gets from the sign it's in, the house it's in, and the aspects it's receiving from the planets and angles. The sign position describes your energy level and, to some extent, how you control and channel your energy. The house position indicates where a great deal of your physical activity will take place and what it will involve. The aspects show particular aptitudes and/or difficulties that will help you or hinder you as you go through life.

In the following chapters, we'll look at Mars by sign, by house, and by aspect, to see its effect on your personality. We'll also look at Mars as a factor in compatibility, occupational success, the best location for you, and many other specific areas of your life. I hope you'll find the material presented informative and that it will aid you in understanding yourself-and others—better.

Introduction to the Revised Edition

It is a rather strange experience rereading something that was written nearly twenty-five years ago. It's also a bit scary. Am I going to have to eat a lot of words? How is this material holding up? Would I still say what I said back then? For the most part, I am happy with what I'm finding. Most of the revisions in this edition are minor and reflect changes in semantics more than anything else. Under the heading of "Occupational Motivation" I have added a brief discussion of Michel Gauquelin's findings on Mars. I have expanded the discussion of intercepted and retrograde Mars. I've also added a chapter on Mars Return charts. And last but not least, the chapters on esoteric Mars have been revised to reflect my current work and line of thinking.

I hope the material presented here will hold up for another twenty-five years. I have had a great deal of pleasure writing and otherwise presenting it over the years, and I hope you will gain an equal measure of enjoyment from learning more about Mars.

Donna Van Toen
Toronto, Canada
April 2011

Chapter 1

Everyday Mars

Before getting into specific areas of action, a general look at Mars is in order. This will tell you what role Mars plays in shaping your personality, what traits it adds to your makeup. It will also tell you what types of energy predominate in your life and what motivates you to take action.

There are four main types of energy: 1) physical energy, which is expended in sports, physical work, and pursuing what you want in life; 2) practical energy, which is expended in fulfilling responsibilities, maintaining a job, and getting ahead materially; 3) mental energy, which is used for mental work, thinking, and communication; and 4) emotional energy, which is used in the sexual area, in feeling, and in protecting yourself.

Mars in the Signs

In terms of what motivates you to take action, the sign and house position of Mars hold the clues. These motivating factors are, in effect, what turns you on! The following pages outline some basic interpretations for Mars, based on its sign, house, and aspect positions.

Mars in Aries

Energy levels: All high—practical, emotional, mental, and physical.
Type of energy: Primarily physical, raw, and spontaneous.
Energy often channeled into: Leading, starting things, pioneering.
Strengths: Courageous, independent, enterprising.
Lacks: Humility, patience, self-control.

This is a powerful Mars in that it provides a great deal of energy, but unfortunately there's often a lack of stick-to-itiveness. You're inclined to leap before you look, with the result that you often find you've got too many irons in the fire to successfully control. There's a hot temper here, although normally there's no tendency to violence. However, if Mars is under heavy stress —in a Grand Cross for example—you could be a heavy drinker or a brawler. This Mars gives a slight tendency to attract cuts and burns, which of course would be more likely if Mars is under stress.

You have a healthy ego and a strong but rather hair-triggered sex drive. You are spontaneous and easily aroused as a rule. Men with this placement are often incorrigible women-watchers regardless of their romantic status and their happiness (or lack of happiness) with same. They find the female body a turn-on. Alas, they sometimes forget that love is not a ninety-yard dash, but rather a cooperative event, and there is generally no applause for coming in first! Women with this placement generally enjoy and are active participants in sex. They are also relatively creative sexually.

Mars in Taurus

Energy levels: High to adequate practical and emotional energies; low mental and physical energies.

Type of energy: Stable, controlled, persistent.

Energy often channeled into: Keeping things as they are, building of both literal and figurative types, accumulating things.

Strengths: Stability, craftsmanship, loyalty.

Lacks: Flexibility, inner drive, physical energy.

You're much slower to act and much more patient that someone with Mars in Aries. Anger can be a problem for you. You tend to hold it in for too long, so when you do blow up it's often over something trivial and your anger is totally out of proportion to the circumstances. You're otherwise quite conservative in your actions and ambitions. In the health area, there's some danger of throat problems, particularly if Mars is under stress or in the sixth house.

You have a powerful sex drive and can be quite sensuous and earthy. Men with this placement have a tendency to want their women to feel obligated to them; they therefore aim to please and also often aim to procreate. Women with this placement tend to deal easily with the requirements of marriage and family life, including sex. They do, however, take a while to "warm up" and don't like to be rushed into lovemaking.

Mars in Gemini

Energy levels: High mental energies; adequate physical energies; low practical energies; emotional energies vary from high to low and are generally erratic at best.

Type of energy: Variable, often fluctuating, nervous.

Energy often channeled into: Learning, talking, writing.

Strengths: Alertness, wit, spontaneity.

Lacks: Depth, follow-through, calm.

Your fondness for change may make you want to do a lot of traveling. Or maybe you'll be the eternal student, constantly reading or taking courses. You're no doubt a clever person, with a vivid imagination and good manual dexterity. But you need a focal point in your life in order to channel your energies constructively rather than wasting them in dabbling. If this Mars is under heavy stress—in a Grand Cross, for example—you can be overly excitable.

Sexually, you fluctuate; you can be very involved one minute, totally superficial the next. Men, when "in the mood," have all the necessary attributes to be accomplished and utterly devastating lovers when they take the initiative; however, those old jokes about Gemini being prone to answer the phone in the middle of an encounter can have a grain of truth! These men can be easily distracted. Women with this placement tend to be turned on by intellect, sense of humor, and other more or less mental attributes. If these are lacking, neither phenomenal good looks nor an evening in the finest restaurant will kindle the fires. People with Mars in Gemini are not insensitive; they have a great desire to learn what pleases. Once they know what pleases, they will generally adapt quickly to the preferred approach.

Mars in Cancer

Energy levels: High emotional energies; adequate to low physical and practical energies; low mental energies.

Type of energy: Predominantly emotional, sympathetic, and subtle.

Energy often channeled into: Gaining security, patriotism, domestic activities.

Strengths: Ability to give TLC, instincts, feelings.

Lacks: Perspective, direction, spontaneity.

The fighting side of Mars is subdued here. You fight openly only when forced to do so; you prefer to use guilt to keep things under control, or you will sidle into things on the defensive, wondering if you'll be able to take the heat or be driven out of the kitchen by more assertive minds-bodies. I call this the mark of the worrywart as your strong needs for security and nice things can put a lot of pressure on you to be "the best" in some way—and of course perfection is hard to come by! Sometimes this position of Mars indicates difficulties in the involved house, because Mars here is in its fall and therefore hard to channel constructively. Other times, since Cancer is associated with home and family, it can attract a great deal of domestic discord. Your intuition is powerful; you should use it. But guard against negative psychism and round out your development by also taking courses in philosophy or theosophy, to aid your psychic development. With practice, you could have good results with astral projection, past-life recall, and tarot, in particular.

Mars in Cancer tends to have a somewhat uneven or erratic sexual quality. People with this placement can go for lengthy periods when they have little or no interest in sex. These periods are interspersed with other periods of extremely sensuous, amorous, and occasionally even promiscuous behavior. Sometimes, there are bouts of impotence or other sexual problems. However, if you're sure you're loved, you're liable to be both refined and faithful, even if not always in the mood for sexual frolics.

Mars in Cancer men have many appealing qualities, not the least of which is the fact that they like women and are protective and solicitous of them. They make marvelous friends, and if you're willing to settle for cuddles instead of fireworks when they're in their disinclined stage, they can be quite capable of meeting your needs. Mars in Cancer women often have some difficulty deciding whether they want to be Earth Mother or Lolita—and sometimes take turns playing both roles.

Mars in Leo

Energy levels: High physical energies; adequate practical and mental energies; adequate to low emotional energies.

Type of energy: Confident, exuberant, dramatic.

Energy often channeled into: Getting others to do what you want them to do, being in charge, entertaining.

Strengths: Generosity, leadership ability, enthusiasm.

Lacks: Ability to work without recognition, humility, ability to play second fiddle.

You're an inspirational person and probably have a considerable amount of self-confidence. Your willpower is also strong. You have a quick but easily-appeased temper. Normally you don't hold grudges. Sometimes you can be a bit too pushy, though, and if Mars is under stress you must watch out for a tendency to be domineering. (On the other hand, if Mars is in a Yod, you have to watch out for domineering people.)

You have a direct approach when it comes to sex and you want what you want when you want it. Men with Mars in Leo are normally optimistic about their chances of scoring and have tons of ideas for the creative pursuit of the opposite sex. Women with this placement can appear haughty. They're generally not looking for just any old passion; they want THE grand passion. Hence they dig their heels in and it takes some effort to sweep them off their feet. People with Mars in Leo generally have a tremendous amount of sex appeal even if they are not traditionally good-looking.

Mars in Virgo

Energy levels: High practical energies; adequate to high mental energies; adequate physical energies; low emotional energies.

Type of energy: Service-oriented, disciplined, efficient.

Energy often channeled into: Reasoning, finding flaws, helping others.

Strengths: Self-discipline, technical aptitudes, unselfishness.

Lacks: Ability to see the forest for the trees, imagination, enthusiasm.

You can be a bit like the Boy Scout who drags the little old lady across the street in spite of her protests that she doesn't want to go. You mean well, and probably wouldn't push people around or argue with them for your own sake, but you have awfully definite ideas about what's good and bad for people in general and those you care about in particular. As a result, you can be too critical of others, interfering more than is good for either them or you. This trait can be channeled positively, however; for example, your sort of motivation can be a good asset if you're involved in a career in the medical field, clerical work, computer programming, or statistical work. In any case, this Mars can be best used for acquiring knowledge and putting it to practical use. With this position, the enthusiasm of Mars is dulled to a large extent; although you tend to be physically active, you're not keen on strenuous activity unless there's some practical purpose to it.

Mars in Virgo people generally tend to be very careful and discreet about sex. Occasionally they're late bloomers in this area. Various types of sexual hang-ups can manifest if Mars is under stress here, and usually have to do with rejecting people for rather petty reasons. ("But I couldn't possibly sleep with him; he'll be bald by the time he's 30!") Both sexes have to learn that good reasoning skills don't make for infallible judgment in the sexual arena. Men with this placement are often so busy repressing their emotions that they may not be very insightful about a woman's wants—or their own. Women with this placement set high standards for men—sometimes unattainably so. Both sexes can be incredibly dense when it comes to picking up on someone's romantic signals; paradoxically, they find very blatant overtures offensive. Patience, as well as more than a modicum of refinement, will be necessary to break the ice—which generally turns out not to be ice at all, but rather a protective coating of intellectual polyurethane designed to protect the emotions from getting scratched, scuffed or stained.

Mars in Libra

Energy levels: High mental energies; adequate to low physical energies; low practical and emotional energies.

Type of energy: Pleasant, relaxed, social.

Energy often channeled into: Socializing, diplomacy, mediating.

Strengths: Fairness, ability to compromise, cooperativeness.

Lacks: Decisiveness, independence, loyalty.

You have a strong sense of fairness, but because you're not always as perceptive as you should be, you may often find yourself torn between two choices of action. Sometimes this Mars position signifies a great deal of inner conflict centering around opposing needs for independence and supportive people. In any case, much of your energy is channeled into partnerships of various types that may or may not be long-lasting. Your physical energy level tends to fluctuate, and a lot depends on the aspects you have to Mars. Aspects between Mars in this position and Moon, Venus, or Neptune can make you rather lazy, while aspects to the Sun, Jupiter, and Pluto tend to increase physical energy.

I have often suspected that a certain number of people with Mars in Libra prefer cocktail parties and pleasant social chit-chat to sex. Sex, after all, is so doggoned messy! It can really play havoc with fancy hairdos, and afterwards there's inevitably all that dirty laundry. And for what?

People with Mars in Libra have no qualms about initiating sexual encounters and no qualms about walking away from them when they cease to please. They don't, however, walk away until they have somewhere else to go as a rule. Men with this placement tend not to be sexually demanding; nor are they anti-sex. They can take it or leave it, depending on who's offering it. Women with this placement are most responsive—and most faithful—to lovers who mirror their own personalities. In other words, they tend to be drawn to the concept of total union, and sex is only one small facet of this merging.

Mars in Scorpio

Energy levels: High in all areas—mental, physical energies, practical, and emotional.

Type of energy: Gutsy, probing, non-stop.

Energy often channeled into: Understanding things, being the power behind the throne, finding hidden things of both a literal and figurative nature.

Strengths: Purposeful, perceptive, dedicated.

Lacks: Forgiveness, openness, adaptability.

This Mars gives you tremendous potential to do good or harm, so strength of character is required if you want to channel your energy positively. You have such strong likes and dislikes that the maxim "thoughts are things" bears especially strong meaning. With this Mars, a negative thought about someone else will have tremendous power and will be felt by the person you're thinking about, unless he or she is totally insensitive. Fortunately, the same holds true of your positive thoughts. Needless to say, your emotions are intense, and as this is a position of extremes, you can be a strong ally or a dangerous enemy, and love—if soured—can turn to hate.

Your intuition is strong, your psychic potential good. In particular, you have good aptitude for alchemy, magic, and creative visualization.

Sex and Scorpio are well-nigh synonymous. However, that insatiably voracious sexual appetite is often only a mask for what Mars in Scorpio really wants, which is understanding. For someone to take the time to get close enough to see, touch, and understand the real person on all levels—physically, mentally, and emotionally—this is the ultimate turn-on for Mars in Scorpio. The males, given this kind of encouragement, will be totally faithful and value the woman who succeeds in this regard above all else. Women with this placement find intelligence a turn-on. They especially find the sort of intelligence known as street-smarts to be a turn-on. But they, too, are ultimately looking for understanding above all else.

Mars in Sagittarius

Energy levels: Physical energies high; mental and emotional energies adequate; practical energies low.

Type of energy: Freedom-seeking, erratic, speculative.

Energy often channeled into: Gambling of a literal or figurative nature, publishing, giving advice to others.

Strengths: Hope, optimism, enthusiasm.

Lacks: Discrimination, endurance, consistency.

You have a strong sense of justice and an inner strength based on a positive philosophy of life. You're also courageous and very good at debating, persuading, and popularizing. Your spontaneity is a plus, although at times it can contribute to sloppiness or slapdash work. And should your Mars be under severe stress—in a Grand Cross, for example—your tendency to act without thinking can get you into a fair amount of difficulty.

Sexually, you're expansive and undoubtedly not the sort to "save yourself for marriage." You tend to approach any kind of sexual encounter optimistically, and sometimes with not a heck of a lot of discrimination. If it's offered, after all, it must be a sign that somebody "up there" thinks you need it. If it's not offered, men with Mars in Sagittarius can be quite bold, blunt, and even crude about asking for it. And they have few qualms about one-night or even one-hour stands. What they do have qualms about is committing and taking responsibility; they are not, by nature, Boy Scouts, and hence do not always come prepared. Women with this Mars placement tend not to be really good at feminine wiles and often wind up being the best friend rather than the lover. They do, however, enjoy sex—although they sometimes, to their detriment, get it mixed up with love.

Mars in Capricorn

Energy levels: High practical energies; adequate physical and mental energies; adequate to low emotional energies.

Type of energy: Predominantly practical, attainment-oriented, sustained.

Energy often channeled into: Planning, overcoming feelings of being unloved, organizing.

Strengths: Respectful, ambitious, responsible.

Lacks: Friendliness, sense of humor, humility.

You probably started planning your career while still in your teens—or even before—because you have a tremendous amount of ambition. You like to be in charge, too. Although you have quite a bit of magnetism, you're not a particularly warm person and can be quite aloof for a long time while you're seeing if others measure up to your standards. And when angry, your coldness is enough to freeze everything in your path. You're well-coordinated as a rule, and not averse to physical activity and hard work.

You have a strong—but often well-hidden—sex drive, and can be an excellent lover once you finally trust your partner—and yourself. You tend to have very specific sexual wants and sometimes aren't terribly comfortable with these. You're also very demanding of yourself in terms of putting your partner's satisfaction before your own.

One of the lessons Mars in Capricorn men seem to be forced to learn at a young age is how to be a good sport or a good loser. Carried into the sexual arena, this means they will generally yield quite politely and agreeably to a woman's demands. However, if this doesn't give satisfaction for them, they may start to be resentful and, instead of speaking up (real men are *always* satisfied if they satisfy the woman), they may start to seek out women they can use for sex and sex only. Which usually only serves to compound the problem by making them feel guilty—which makes them try even harder to please their significant other at the expense of their own needs.

Women with Mars in Capricorn tend to be logical about sex. They see it as a need that can be attained. They don't generally confuse sex with love, and they tend to be "good shoppers" who know their rights. All of which sounds terribly cold and unromantic. Unromantic, maybe, but these women—once committed—tend to be both very warm and very wise.

Mars in Aquarius

Energy levels: High mental energies; adequate practical energies; adequate to low emotional energies; low physical energies.

Type of energy: Mentally-dynamic, truth-seeking, scientific.

Energy often channeled into: Avoiding restrictive relationships and possessive people, discovering truth, humanitarian causes.

Strengths: Friendliness, originality, independence.

Lacks: Conformity, empathy, respect for tradition.

Mars is at its best when it has a physical outlet. Aquarius is an intellectual sign and loathes to expend any more physical energy than it has to. Consequently, Mars tends to be uncomfortable in Aquarius and may not be as productive here as it is in some other placings. If you're convinced of the importance of a cause or course of action, there's no problem, as this gives you the motivation to take physical action. Otherwise the energy of Mars can get bottled up inside where it creates inner conflicts and nervous tension. The source of conflict is very often a conflict between your need for friends and your need to be an individual in the truest sense. The latter need tends to make you scornful of things like tradition and conservatism, and hence makes some people a little wary of you, which of course makes you uncomfortable and makes your inner conflict more pronounced. Often you can do well in areas like electronics, science, and writing, where your originality will be appreciated rather than frowned upon.

Sexually, you're experimental and prefer to play the field for a while before settling down. Men in particular find it easier to have women as friends than to have them as lovers—they don't always have an easy time keeping lovers happy. Though kind, they tend to be brotherly rather than loverly and can have a detached, almost clinical, attitude toward sex that can earn them an undeserved reputation for cold-heartedness. These men may need to realize that while sex may well be just another normal bodily function, most women prefer it to have a little more mystique than, say, flossing one's teeth or showering. Women with this placement tend to be turned on by intuition, truthfulness, and helpfulness. Mental orgasms are just as important—if not more so—to this woman as physical ones.

Mars in Pisces

Energy levels: High but erratic emotional energies; adequate mental energies; adequate to low physical energies; low practical energies.

Type of energy: Unifying, erratic, intuitive.

Energy often channeled into: Serving others, unifying ideas or principles, looking for solutions to difficult problems.

Strengths: Charitable, receptive, imaginative.

Lacks: Consistency, drive, goals.

There are plusses to this Mars, but overall it's a difficult placement. On a positive note, this position will help offset selfishness found elsewhere in the chart. It also makes you kind, forgiving, and a good listener. Usually it suggests shyness; people with this position are almost never aggressive and are frequently of the sort who apologize when someone else steps on their feet. But the potential problems are many. Sometimes there's a tendency to be too influenced by others.

Other times there's a tendency to wander through life with no goals. In still other instances, there's a fear of the unknown in the material world that leads to involvement in religion or some form of psychic phenomena to the exclusion of all else. The emotions are almost always unpredictable, with ecstatic highs and a tendency to brood when depressed. This latter tendency needs watching; if carried to extremes it can make you quite neurotic or even trigger a persecution complex. Should this Mars be in a T-square, Grand Cross, or otherwise under severe stress, extra care is needed with alcohol, tranquilizers, and psychic phenomena of the spiritualistic nature, as these can cause you to lose touch with reality.

Sexually? Well, Mars in Pisces can usually take sex or leave it alone. If it's a total union—spiritually, intellectually, and emotionally—they'll take it. If someone else needs it, they'll offer it —particularly if the "needer" is a wounded bird—non-orgasmic, impotent, not awfully confident of his/her own sexuality, for example. Mostly, though, Mars in Pisces is just as happy merely being held as he or she is being ravished. Mars in Pisces men are phenomenally patient; they therefore often attract women with various sorts of sexual hang-ups. Often they turn out to be just what the doctor ordered for these women—although occasionally they'll need a doctor for themselves by the time the affair's over! Women with Mars in Pisces tend to be somewhat leery of closeness, as if they fear losing themselves and their own identities in the relationship. They often prefer friends to lovers for this reason, and invariably go through a phase of being terrified to discover that there are sexual feelings in the relationship. The terror is worse if she can trace these feelings within herself.

However, once this phase is transcended (which it can be), this woman is loyal and loving, though she may always prefer the after-cuddles to the actual sex.

Mars in the Houses

While the sign position of Mars shows the primary type of energy available to you, the house position shows you the primary area of focus for that energy. In other words, this is going to be one of the most active departments of your life, and one in which you're highly motivated to show the world who you are and how well you can stand on your own two feet. Or, if your Mars is uncomfortable here, this may be a place where you frequently feel threatened and are pushed by events or by your own feelings to defend yourself and anything here that you identify with as being intrinsic to what you are.

Mars in the First House
Courageous, impatient, and sometimes pushy, you don't always take the time to understand other people's points of view. So although your leadership abilities are good, you may drive your followers, employees, et cetera, a bit too hard. Sometimes this gives red hair. Or there may be a scar on the face. Stress aspects to this Mars tend to bring high fevers when ill.

Mars in the Second House

You're efficient, acquisitive, and fairly possessive. But you're not tight with your money. You spend freely and can at times be very extravagant. You could have a loud voice. If this Mars is under heavy stress—such as in a Grand Cross or T-square—there can be losses resulting from fire or theft.

Mars in the Third House

When you're talking with people, be sure to keep your impulsive streak under control; otherwise, you may find yourself having to eat your words fairly often. You're frank, determined, and sometimes pushy. You're also a risk-taker. If Mars makes any aspect to Uranus, fear isn't in your vocabulary. You'll fight for yourself and your beliefs. You'll also fight on behalf of your neighborhood and family. But if neighbors, siblings, cousins, and the like disagree with you, you can just as easily fight with them instead of for them. Mars here as part of a Yod indicates dealings with extremely excitable, possibly unstable, people on a daily basis.

Mars in the Fourth House

You may have sudden and inexplicable touchy spells. You're a hard worker in your home and/or for your loved ones, and are normally energetic (although if Mars is in a water sign your energy may be more emotional than physical). Possibly you lost a parent in childhood or came from a home where fighting was common. Early on in your life, strong desires for independence and security came to the fore. These are what motivate you and no doubt what you gain in these areas will be retained well into old age and perhaps throughout your life. Sometimes people with this aspect prefer not to marry, but they nearly always want their own home. Mars here under stress can be a troublemaker, a disturber of the status quo. Watch that!

Mars in the Fifth House

Creative and passionate, when you meet someone you like, you probably won't take no for an answer! There'll probably be lots of romances in your life for that reason. You should be able to work well with children; you may also have aptitudes for acting or dancing. If you have children of your own, they'll probably be a lively bunch, and there's a good chance the first-born will be a male. One of your children is apt to be extremely stubborn and may cause some unhappiness in your life.

Mars in the Sixth House

When working with others, you're inclined to be bossy, but as you're a hard worker, you'll be respected if not always liked. If you employ others, you'll tend to attract energetic workers. You have a tremendous amount of vitality if Mars is in a fire or earth sign, and even in air and water signs this placement can increase physical energy a bit. You may suffer from headaches as a result of this Mars, especially if it's under stress. And if Mars is part of a Grand Cross, you may overwork to the point where your health is adversely affected.

Mars in the Seventh House

This suggests you have some lessons to learn about partnerships in general and romantic commitments in particular. Public opinion will also somehow play a role in your life. Highly independent and courageous yourself, you're apt to attract an energetic, assertive mate. There may be a bit more competition than is healthy between you, particularly if you're both employed. Enemies tend to be aggressive, selfish, and highly competitive people.

Mars in the Eighth House

Lusty, blunt, and an independent thinker, you do things with a great deal of intensity. You tend to live for today and aren't especially good at saving, so it's possible there will be a great deal of fighting over joint finances in your life. You have strong sex drives and will try to be the dominant sexual partner. In the career area, you may be attracted to surgery, psychology, or psychiatry. Your death is apt to be connected with some sort of accident. Although your psychic ability is good, if there are stress aspects between Mars and Neptune, you should avoid involvement with psychic phenomena due to a tendency to become fanatical about it.

Mars in the Ninth House

This says you're quite restless. Mars here increases your mental energy, but also broadens your interests—sometimes to the point where you become a dabbler rather than capitalizing on your best potentials. There may be fights with your in-laws. In rare cases, there are ethnic or racial prejudices. If Mars is in a Grand Cross or otherwise under severe stress, there can be bad timing in espousing causes, and poor choices in terms of leadership.

Mars in the tenth house

This position suggests several kinds of job aptitudes: metalworking, mathematics, politics, unions, or military, or osteopathy, chiropractic, or other alternative healing. Argumentative and independent, you want to be one of life's winners and are ambitious to reach the top in your career area. And if you're in the public eye, you're highly motivated to make a good impression.

Mars in the Eleventh House

The energy and motivation of Mars is expressed through your life goals and your powerful dreams for the future. An enterprising person, much of your energy goes into creative activities. You have many acquaintances, the majority of whom are male. If Mars is in difficult aspect to other planets, planning is not your strong point and you have a tendency to leap before you look.

Mars in the Twelfth House

A deep, secretive person, you tend to be more or less clueless about how you affect others. You have a strong impulsive streak and, generally, a tendency to buck authority and go against tradition. This brings a danger of slander into your life, as it makes you quite a controversial person. If Mars is in a Grand Cross or otherwise under severe stress, you can be extremely disruptive.

Aspects to Mars

Aspects to Mars give us further information about your assertiveness (or lack thereof), your energy level, your motivational vitality, et cetera. If your Mars has hard aspects, you may over-compensate by becoming super-aggressive, or you may decide on retiring yourself from competitive situations entirely. These somehow, depending on the signs, houses, and planets involved, may discourage you from being yourself and can, if not dealt with, lead to wasted energy and wasted potential. On the other hand, if you have a relatively laid-back Mars, say in Pisces or Taurus, these same aspects can spur you to make more use of your energy, to push yourself a bit harder. So it's a mistake to assume that any "hard" aspect is undoubtedly a "bad" aspect.

The sextiles and trines, on the other hand, show the potentials you have for expressing your energies in ways that others will appreciate and respect. These, then, symbolize constructive outlets for your assertiveness and ambitions.

Sun Conjunct Mars
This tends to add energy and make you more fond of physical activity than you might otherwise be. Energetic and competitive, you have a great deal of leadership ability. This does, however, increase the danger of miscarriage in a woman's chart.

Sun Semi-sextile Mars
You tend to be defiant and this creates a problem in the involved houses. You go after what you want in this area regardless of what others think. You don't always succeed, though, as there's often a conflict between your emotions and your ability. A tendency to overwork in the involved area can also cause problems.

Sun Sextile Mars
This is good for the health and increases the energy, which often has an intellectual outlet. Ability for law, journalism, and/or broadcasting is common. Sexual appetites tend to be controlled positively. This is almost always the mark of a courageous person. Occasionally it contributes to recklessness.

Sun Square Mars
This aspect encourages aggressive tendencies. Another thing it can incline you to do is to feel very angry at the slightest frustration. Older books say you should avoid being around guns. If you learn to think before you speak, you could be an excellent debater. You could also do well in the business or medical world; although you may not be popular, you know what it takes to get things done right and will push for that.

Sun Trine Mars

In group activities, you tend to be daring and blunt, but are cooperative as long as you have a say in things. This is usually an indication of some sort of additional strength as well as good brainpower.

Sun Quincunx Mars

You have difficulty handling your anger and tend to hold it in much longer than is good. Impulsive and energetic, you have an enormous amount of ambition, but your tendency to blow up at inappropriate times, over seemingly trivial circumstances, may hold you back. Assertiveness training could help this.

Sun Opposition Mars

Aggressive and quarrelsome, you probably have plenty of courage. You generally get what you want from life, but you don't win too many popularity contests. A tendency to burn the candle at both ends can tax your strength.

Moon Conjunct Mars

Courageous, daring, and sometimes impatient, the main problem to watch out for is a tendency to fight for ideas merely because they're unpopular rather than because they're good. You may be moody. You have strong sex drives. Women with this aspect often espouse feminist views; men tend to be traditionalists. Both tend to alienate the opposite sex at times.

Moon Semi-sextile Mars

Your relationship with your mother has created ambivalence about matters in the involved houses, which has in turn created a conflict between your emotions and your actions. Your mother may have told you, ''Do as I say; not as I do,'' in this area. You're a passionate person. You tend to quarrel and to take unintentional slights personally.

Moon Sextile Mars

Robust, forceful, and physically strong, it's important that those around you give you enough freedom to express your personality. Otherwise, you can develop a "live for today" approach to life that is counterproductive to your goals. Your passions and your resolve are both strong. This aspect can offset other more problematical health aspects.

Moon Square Mars

You tend to leap before you look and to let your feelings overwhelm your logic. You either get on the defensive too quickly yourself, or you attract defensive, contentious people. Either way, you tend to get involved in a lot of emotional crises. A lack of self-respect is common. This aspect is not a plus in terms of health.

Moon Trine Mars

You're an energetic, ambitious, strong, independent individual. You tend to be attracted to people who are just as strong-willed, energetic, and ambitious as you are; you don't like weak, clingy people. Usually this aspect is indicative of robust good health.

Moon Quincunx Mars

Usually this indicates a difficult—even destructive—relationship with your mother. There's a tendency to jump the gun on things and to be too quick to be on the defensive. Your strong emotions may influence you excessively, causing you to feel sorry for yourself. Or there may be a tendency to become involved in other people's problems in order to avoid having to face your own.

Moon Opposition Mars

Headstrong and aggressive, you must develop self-control in order to counteract the quick, fiery temper this gives. Like the square, this aspect is not an asset in terms of health.

Mercury Conjunct Mars

This tends to be easiest to channel when in a mutable sign and/or a cadent house. In fire signs, cardinal signs and/or angular houses, it tends to give you more energy than you know what to do with. In any case, it makes for a practical person with a sharp, quick tongue and mind. You love a good discussion and have ability for debating, journalism, broadcasting, and acting as a commentator. You may also have mechanical ability.

Mercury Semi-sextile Mars

You might find it hard to complain to others or you might spend most of your time complaining to others. Probably when you do complain, you complain to the wrong people (for example, you complain to your friends about your parents, to your parents about your mate, to your mate about your boss, et cetera). You tend to be well-informed but aren't always willing or able to use your information constructively. There's forcefulness here, but it needs constructive channeling. There's also a way with words, but this is sometimes marred by a tendency to exaggerate or poor judgment about whom to talk with.

Mercury Sextile Mars

Good-natured and plain-spoken, you think for yourself. This aspect strengthens your mental energies and encourages you to keep learning throughout your life. It's also said to improve your eyesight. Often people with this aspect choose not to have children.

Mercury Square Mars

When you're angry, you tend to be overly forceful. This normally encourages materialistic tendencies and in extreme cases can contribute to criminal tendencies. More commonly, it gives good aptitudes for law, teaching, lecturing, writing, and/or working with tools.

Mercury Trine Mars

This is a mind-invigorating aspect that sharpens and strengthens your mind, increases your mental energies, and encourages you to keep learning throughout your life. It also encourages self-confidence.

Mercury Quincunx Mars

There's a lack of self-control or moderation in the traits or activities signified by the involved signs and houses. You may be excessively fault-finding, opinionated, or inconsiderate.

Mercury Opposition Mars

This aspect can make you feel nervous or restless much of the time. Irritability needs conscious control; meditation can be of value here. You have a good mind, but can be a bit of a know-it-all, so this isn't exactly a positive aspect in terms of popularity.

Venus Conjunct Mars

Your relationship with your parents will determine your attitudes toward the opposite sex and marriage, for better or worse. Normally this gives a harmonious blending of active and passive energies and makes you sensitive and loving unless both planets are under severe stress. Drive is generally expressed constructively. This aspect is an asset when it comes to relating to the public and is said to give luck in financial matters.

Because this conjunction can vary dramatically from sign to sign, and because it tends to give you valuable insights into your earliest dealings with the opposite sex—including dealings with your opposite-sex parent, siblings, and peers—a sign-by-sign run-down may be useful.

In Aries: The conjunction often suggests that either the parents were just kids themselves when they had you, or parents who tried—successfully or otherwise—to share your youth by keeping interested in "your" fashions, music, etc. You probably started noticing the opposite sex at an early age and virtually every romantic experience you've ever had is indelibly tattooed on your mind, for better or worse. You stride through life bravely and rapidly—so rapidly in fact that others can't always keep up with you.

In Taurus: The conjunction suggests an upbringing that's instilled a considerable ability to enjoy life. Your opposite-sex relationships were probably reasonably stable, and for this reason you have come to expect stability of the opposite sex. You take commitment seriously and woe be unto those around you who commit and then let you down, be it socially, financially, or otherwise!

In Gemini: The conjunction suggests that you may have been brought up to make your own decisions from an early age. Your most fulfilling moments in life are generally connected with learning, writing letters, or talking on the phone. You have learned to make your expectations clear with a few well-chosen words —a skill that should smooth your relationship path.

In Cancer: The conjunction says that security is definitely a big issue with you. Either you didn't have any when you were growing up, or perhaps you were so cherished that no other environment can satisfy you. You can be crabby, contentious, and petulant if you perceive yourself to be in a hostile environment—and any environment you're in for the first time is most assuredly perceived as hostile. However, once you've established a secure relationship in that environment, you can be the gentlest, most tolerant, most giving person anyone would ever want to meet.

In Leo: This is often the sign of a late bloomer. I'm not sure why, unless perhaps it's because these people are such charmers that they can get away with murder long past the stage when most kids are being told to grow up and act their age. They have a childlike radiance that can turn away anger and other "heavy" emotions with astounding facility. People tend to be faithful to these folks for a long time, come what may. Normally this faith is justified, though this conjunction can have a darker self-serving side: sometimes Mars in Leo uses others as props to draw attention to himself or herself.

In Virgo: The conjunction suggests an upbringing replete with all sorts of rules and rituals and a motto of, "Happiness is finding a good reason for feeling the way you do." Very occasionally sex was a dirty word in the early years. Instead of sexual pleasures, the joys of serving others, working hard, or abstaining from fleshly delights of all sorts may have been stressed. The other notable thing about this placement is that these people often handle criticism much better than praise.

In Libra: The conjunction suggests a family life where people were more interested in their individual social lives than in each other. It could have been a perfectly harmonious environment, but it probably had a certain unreal, "Leave It To Beaver" quality. Raised voices, real anger, and intense emotions weren't visible. So as an adult you find these to be interesting challenges.

In Scorpio: The conjunction suggests a situation that may have been much the reverse of the above. You were probably instilled with a belief that intensity is the key to success and that the purpose in life is understanding why what happens happens. You may have been exposed to some sort of black moods or vindictiveness. You therefore probably grew up to be someone who tends to become totally immersed in an activity. When the time comes that it no longer meets your needs, you discard it. Unlike many, you have the nerve to burn your bridges and face the unknown, for you feel if you have nothing to lose, there's no point in hanging around.

In Sagittarius: The conjunction indicates that you were brought up to refuse to take hard times seriously or give in to depression. The motto for this might be: "Happiness is sharing whatever you have." The more you can partake of gambles, the happier you are, because you *know* that pot of gold is out there at the end of the rainbow and feel the more you explore, the greater your chances of finding it.

In Capricorn: The conjunction indicates that you were brought up to believe that the best things in life last a long time. You therefore do not live for the moment, but rather look at the long term. Often this is a mark of upbringing where "childish" behavior was strongly discouraged. As a result, there's often a very vulnerable child within who has never had a chance to work through earlier "normal" fears and feelings. You may, therefore, look for help in attaining inner peace, fulfillment, and belief in your own lovability.

In Aquarius: The conjunction indicates that you could have been straight-jacketed by well-meaning parents who were totally clueless about your individual wants and preferences. Or, you could be someone who no sooner got comfortable in what you thought was a stable routine when someone walked in and disrupted it. You were instilled with a deep sense of honesty and have no qualms about saying what you think. You are also someone of whom it's best not to ask a question unless you're sure you want to hear the answer. In romantic/ social matters, you're friendly, but will go to great lengths to avoid being possessed or tangled up in other people's lives.

In Pisces: The conjunction suggests an upbringing where physical activity was somehow discouraged. In later years this can translate into a cluelessness about your physical body wants, tolerances, and capabilities. You can therefore become quite consumed by attractions to one thing or another, and can become addicted to another person as easily as others become addicted to alcohol or drugs. Most of the time people with this aspect are at least somewhat addicted to pulling things, as well as people, together. Occasionally they need to be gently reminded that if they don't have their own acts together their effectiveness in other situations is liable to be limited accordingly.

Venus Semi-sextile Mars

You may be attracted to people who are wrong for you for some reason, making it difficult to satisfy your romantic desires. You have warm feelings and sex appeal, but somehow tend to be on the wrong track when it comes to choosing a mate. So while this aspect always increases your chances of marrying, it also increases the risk of eventual separation.

Venus Sextile Mars

You have an even, well-balanced flow of sexual and romantic energies, meaning you're neither overly aggressive nor overly passive in your approach to these areas. Love-hate relationships aren't your style; nor are purely platonic ones. You tend to be happiest with an easygoing mate who has a considerable amount of imagination and is creative without being a moody, broody, artistic archetype. Your attitude toward sex is good overall. You're attracted to marriage and are generally comfortable in your role as a spouse, provided of course that you choose a compatible mate.

Venus Square Mars
Insecure relationships or very difficult partnerships are common. You tend to be flirtatious and can be self-indulgent. Often people with this aspect are quite good-looking. You tend to put too high a value on the trappings of success and sometimes not enough on inner satisfaction.

Venus Trine Mars
Affectionate, pleasure-loving, and passionate, you have an intense approach to life that lasts well into middle age and sometimes longer. You have a great deal of warmth, which attracts others to you.

Venus Quincunx Mars
You may be attracted to people who are wrong for you romantically, making it difficult to satisfy your desires. You're often discontented without knowing exactly why. You can be passionate to the point of dissipation. You can also be too ambitious for your own good.

Venus Opposition Mars
Either your relationship with one or both parents was difficult or their relationship with each other was difficult. Impulsive and sometimes fickle, you tend to have a rather turbulent love life. Hypersensitivity is common when this aspect is present.

Mars Conjunct Jupiter
An enterprising person, you like to be the first in your circle to try new things. You have a great deal of energy; in fact, the biggest problem with this aspect tends to be physical exhaustion from driving yourself too hard for too long. Often this aspect greatly enhances earning potential. There may be involvement with teaching at the elementary, high school, or college level or with the military. If in a fire sign, this aspect can make you overly expansive, aggressive, or reckless.

Mars Semi-sextile Jupiter
You think you're unlucky when in reality you're wasteful and tend not to answer the door when opportunity knocks. A sort of rebellious bravado is common. There may also be a problem with your temper. You tend to take on more than you can handle. Feast or famine is a common theme.

Mars Sextile Jupiter
This increases your energy level. Normally the wisdom of Jupiter directs the energy of Mars constructively. An enthusiastic person, you tend to be very secure in your own beliefs. Both travel and friends tend to be very beneficial in your life.

Mars Square Jupiter
In spite of your reckless streak, most people like you and find your generosity and your high energy level to be attractive qualities. But on a personal level, your restlessness can cause you seri-

ous problems. You tend to overdo things and therefore need to make sure that your goals are well-defined. This aspect can contribute toward a tendency for high blood pressure and stroke.

Mars Trine Jupiter
You have willpower, optimism, and enthusiasm, and dearly love your freedom. This is a super-powerful energizing aspect.

Mars Quincunx Jupiter
You have a tremendous need for freedom and were probably difficult to discipline as a kid. An excitable, overly-enthusiastic person, you may be victimized by people who take advantage of your willingness to become involved in noble but unrealistic goals and causes.

Mars Opposition Jupiter
Hastiness and extravagance are common with this aspect, so self-discipline is absolutely necessary. You detest almost any kind of authority. Sometimes there's a tendency to be accident-prone.

Mars Conjunct Saturn
Even at best, you're apt to feel frustrated fairly often due to a conflict between your inhibitions and your motivations. You don't lack courage, but your self-confidence tends not to be very high, so you sometimes feel you can't do things when you really can. You could make a successful career of the military or politics; you may also have good aptitude for math. There may be accidents that cause broken bones. This is especially true if Mars or Saturn is under stress, as this tends to encourage recklessness.

Mars Semi-sextile Saturn
Sometimes there's someone in your life who discourages you constantly or makes it hard for you to determine where your responsibilities lie. You tend to be brave one minute, afraid the next. This aspect can contribute to intolerance and, at worst, erode your sense of purpose.

Mars Sextile Saturn
Although you are courageous and persistent, there's almost always a definite problem when it comes to constructively handling your anger as you tend to turn it inward more than is good. You have a considerable amount of organizational ability. Your home and career tend to be very important to you.

Mars Square Saturn
Fights of a verbal nature are pretty common, and in extreme cases there's a possibility of some physical violence in your life. Sometimes there's merely a tendency to break bones more easily than the average person. In general, there's a lack of enthusiasm for life and a tendency to be

ambivalent about things. Your early life may have been hard, leading to an erratic energy flow or apathy in adult years.

Mars Trine Saturn

You have endurance and self-assurance and are a meticulous worker. You're especially good at work that requires attention to detail. You tend to lead an organized, productive life. This aspect also increases your magnetism.

Mars Quincunx Saturn

On the plus side, you're generally not aggressive; on the negative side, there's normally a lot of resentment connected with the involved houses and what they denote. Your energy flow is erratic; you alternate between energetic activity and heavy apathy. You tend to either erroneously assume that others are more talented than you or vice versa. You tend not to be completely honest with yourself for one reason or another.

Mars Opposition Saturn

You find it difficult to maintain an even flow of energy and tend to be ambivalent about things, although if you discipline yourself you can be good at detail work, particularly where crafts are concerned. You can be overly aggressive or passive-aggressive. In extreme cases, this aspect can attract violence.

Mars Conjunct Uranus

Courageous to the point of fearlessness, you sometimes tend to be against certain people or ideas just because other people are for them. Should either Mars or Uranus be under severe stress, you could be extremely headstrong and hard to control. You may be mechanically inclined or have a special ability for science or writing.

Mars Semi-sextile Uranus

You need to find an outlet for your energy—preferably something connected with the involved houses. Otherwise you'll feel pressured. You dislike routine and can become scatterbrained as a result of boredom. Your desire for freedom borders on the obsessive.

Mars Sextile Uranus

When others try to pressure you to follow a course of their choice, you tend to behave impulsively and do the reverse of what they want. You can be forceful, but sometimes pay a price for this in terms of nervous strain. You have good intuition, which can serve you well in the business world, or as a sociologist, social worker, or astrologer.

Mars Square Uranus

An irritable person and a fighter, your impulsive behavior can have a negative impact on your health. Nervous tension may be a problem. Or you may be accident prone. There's a tendency toward recklessness in the romantic/sexual area and choice in friends is not always the best.

Mars Trine Uranus
Because you are energetic and daring, people will probably like your style. You're able to make quick decisions and may be a bit of a reformer. This aspect tends to increase physical strength.

Mars Quincunx Uranus
Sometimes you make sudden decisions or take sudden action in a way that startles others. Hence, you're vulnerable to public criticism. You're impulsive to the point where your actions are sometimes self-defeating. One positive outlet for your impulsiveness may be in the area of humanitarian causes or work for the underprivileged.

Mars Opposition Uranus
Because you are an argumentative person, there's a danger that your behavior may lead to accidents or, in extreme cases, violence. You want to go your own way and may be a bit of a revolutionary.

Mars Conjunct Neptune
Psychic and perceptive, you have a powerful imagination and can at times be full of hot air. You no doubt have a dramatic streak. Your sense of responsibility may not be as strong as it should be. Occasionally this aspect has an adverse effect on the health, with allergies or a low physical energy level being the most common complaints. Mars conjunction Neptune rarely does wonders for self-confidence. Some people with this placement are inclined to bolster shaky self-esteem with a drink, a toke, or a snort to help them cope with social situations, presentations, etc. While most remain social drinkers (or smokers or whatever), it is not unknown for those with this aspect to occasionally slip over the edge and into dependency on one of these crutches. Generally though, the most serious problems with alcohol and drugs come from the opposition.

Mars Semi-sextile Neptune
Lack of energy or misuse of energy is common. A positive outlet is needed for escapist tendencies; this could be something like acting or it could be something creative linked to the involved house or houses. There can be a tendency to exaggerate. There can also be coarseness or sexual hang-ups.

Mars Sextile Neptune
Your approach to sex is forthright yet low-key. You're able to size up romantic opportunities fairly accurately and are able to tune in to people's needs and respond accordingly. You're warm without being smothery. You're therefore able to handle those difficult, strong, silent types or passive, shrinking violets better than the average person. You're also sensitive to the

needs of those who aren't able to take care of themselves, which is fine as long as you don't constantly find yourself attracted to weak, helpless types and stuck in the role of martyr as a result. You tend to be attracted to sympathetic, rather complicated personalities and would have a special attraction to those involved in the helping professions or in creative fields.

Mars Square Neptune

This tends to weaken you or decrease your physical energy level. Worry can affect your health adversely. You're also susceptible to allergies and food poisoning. You must learn to use your logic to keep your imagination in line; otherwise, obsession and/or depression can occur.

Mars Trine Neptune

A sympathetic, enthusiastic, creative person, your feelings are nearly always aroused when you see people who seem helpless or lost. Your emotions are extremely powerful and should be channeled into creative work for best results.

Mars Quincunx Neptune

Physically, you may not be strong and you tend to not get much in the way of regular exercise. There's a tendency to be attracted to drugs as magic cure-alls—not just the mood-altering ones—but things like hair restorers and weight-loss nostrums. Often your imagination is too sensitive; in extreme cases, this can lead to a morbid streak. In the relationship area, you tend to leave yourself open to deception because you expect the best of everyone and tend to ignore evidence that they might be anything less than perfect.

Mars Opposition Neptune

Sometimes this aspect creates health problems that have to be treated by drugs, which can in turn have side effects that adversely affect the health. In extreme cases there's a tendency toward negative escapism through drugs or alcohol. Lack of moderation is common.

Mars Conjunct Pluto

On the plus side, you're adventurous and have very strong willpower. On the negative side, you're highly emotional with a quick temper and wild passions. You won't tolerate too much interference in your life, for even when the energy of this aspect is expressed constructively you tend to be argumentative. If either Mars or Pluto is under severe stress, this aspect can contribute to mental illness.

Mars Semi-sextile Pluto

Either you're aggressive and foolhardy yourself or you attract people of this type. Your goals can become obsessions in any case. If the eighth house is involved, quarrels, criticism, and/or rebellion are apt to be themes in your life.

Mars Sextile Pluto

Courageous and rather unbending, your determined approach to life can have a great many positive results. Normally this makes you a hard worker, someone filled with a burning ambition to succeed. It also strengthens your constitution.

Mars Square Pluto

An impulsive person, you have great ambitions for your life. If you develop the higher vibrations of Pluto, such as analytical skills, discernment, and subtlety, your ability to overcome obstacles will be good. Your strong sex drive will require a constructive emotional outlet.

Mars Trine Pluto

You're full of motivation and ambition and relentless in the pursuit of your goals. This aspect adds both physical and emotional energy and normally helps you to positively channel your energy.

Mars Quincunx Pluto

At certain times in your life there will be some sort of hard or distasteful work to be done and you'll be able to accomplish very little else until this work is completed. You have the ability to overcome obstacles, but a tendency to let others make too many demands on your time needs watching. You can be accident-prone. In extreme cases this can contribute to a violent streak.

Mars Opposition Pluto

You're very ambitious and very forceful. In extreme cases, your goals can become obsessions and you can become a tyrant. If your energy isn't channeled constructively into work or studies that aid self-understanding, it can contribute to a tendency to attract violence.

Mars Conjunct Ascendant

You're energetic, impatient, and very competitive in most areas of your life. You're also passionate in most relationships. You may have red highlights in your hair or there could be a scar or birthmark on your face. If Mars is in the twelfth house, your energy may be harder to channel than with Mars in the first, but at the same time, this gives you the ability to turn what at first seems to be bad luck to your advantage. There may also be an inclination toward psychic development with this position.

Mars Semi-sextile Ascendant

You find it difficult to compromise in close relationships, and every effort must be made to combat a tendency toward the conceit inherent in this aspect. You tend to be an impulse-buyer if Mars is in the second house. Although you often give the impression of being solid and plodding, there's sometimes a great deal of intensity hidden in your make-up, particularly if Mars is in the twelfth house. If this isn't channeled properly, it can cause occasional physical problems.

Mars Sextile Ascendant
A lively person, you like to be free to do what you want. You don't like others to make demands on you, so occasionally there may be arguments with siblings, friends, or relatives. You're quick-thinking, but your follow-through on ideas often leaves something to be desired.

Mars Square Ascendant
You tend to act without thinking when you're feeling angry, sometimes going to such great lengths that you wind up cutting off your own nose to spite your face. It's important for you to learn the art of compromise so that you can constructively combine your energy with other people's instead of worrying about your own needs to the exclusion of everything else. A tendency to be accident-prone needs to be guarded against. A lot of the action in your life will center around your home and work since you take pride in your status and your material circumstances.

Mars Trine Ascendant
A lively, strong-willed person, humility isn't your strong point. You'll defend your beliefs no matter what and won't find it easy to admit you're wrong. Consequently, there may be problems with offspring, in-laws, or those who come from cultural backgrounds that are different from yours.

Mars Quincunx Ascendant
You have difficulty understanding where your rights begin and other people's leave off, or vice versa. Dissatisfaction is common in your life, and non-constructive criticism is also a theme. You put a great deal of energy into your work but are sometimes motivated to concentrate on non-productive work or busy-work to the detriment of your real interests or goals. You may be subject to headaches.

Mars Opposition Ascendant
Either you go looking for fights or you attract people who are looking for fights. In any case, there may be a lot of friction in your life. Enemies tend to be forceful, selfish, and highly visible; you don't attract sneak attacks or backstabbing as a general rule. If Mars is in the seventh house, it increases your energy and drive tremendously, but it also gives a tendency to waver between two equally attractive people or things, making your energy hard to channel. In the sixth house, migraines or tension headaches can be a problem. These are most likely if your daily routine is one where you feel you're existing rather than living, or are stuck in an unfulfilling, meaningless job.

Mars Conjunct Midheaven
You should choose a career in which you can have a great deal of independence. You might possibly have a flair for engineering, mathematics, computer science, or politics. In any case, you

tend to focus on a career where you can gain a position of leadership, and would work for recognition of your achievements. You have plenty of drive and are probably mentally active. In extreme cases, this position attracts scandal.

Mars Semi-sextile Midheaven

Assertive, unruly, and dauntless are all good words for you. Clearly, you have a great deal of energy, and for best results, you should spend a lot of time releasing it. Sports and other vigorous hobbies can be good outlets. Or you might be inclined to join pioneering groups or explore untraveled places. If work is the only outlet for your energies, strain is possible.

Mars Sextile Midheaven

Your relationship with your father is important to your life, for better or worse. This aspect adds diplomacy and inclines you to hold in your resentment. In spite of this, it can attract enemies. Although you're an independent person, goals may be lacking.

Mars Square Midheaven

Reckless, undaunted, and energetic are all good words for you. This aspect seems to make it harder than average to know yourself. Sometimes overwork causes strain. On the other hand, if you feel secure about yourself and your life, this aspect can be a plus as it pushes you to make the most of your potential.

Mars Trine Midheaven

When you have to make a choice, you usually choose what seems to be best for you. Altruism is rarely a strong point; neither is flexibility. This aspect does, however, make you very ambitious for money and generally adds sophistication and a way with words.

Mars Quincunx Midheaven

You're mentally agile but rather indecisive. You don't always say what needs to be said—at least not at the right time! Very little else stops you from getting ahead. Strain resulting from overwork is a possibility.

Mars Opposition Midheaven

Either you don't get along with your parents or you lack self-confidence—possibly both. In any case, strife in your early environment may have caused you to leave home at an early age. Occasionally, parents may have forced you into a career of their choice. In terms of temperament, you tend to be forceful and easily annoyed.

Chapter 2

Mars and Your Love Life

Mars figures heavily in matters of love and lust. Why? Because among other things, it represents sexuality, passion, social assertion, and desire. Yet Mars doesn't stand alone. There's more to love than raw passion and sexual magnetism; there's also a need for sharing, caring, and a certain amount of stability if a relationship is to last. That's where Venus comes in.

Venus represents romanticism, gentleness, femininity, and what you need to maintain harmony. These things are all crucial to a relationship.

So Mars cannot stand alone. It does no good to see how Mars affects your love life unless we consider Venus as well. These are two sides of the same coin. The two sides would ideally make up a whole person, a person who has accepted fully and integrated both the masculine, active, passionate side in his or her nature and the feminine, passive, gentle side. In other words, to fully understand Mars and its role in your love life, we must first look at Venus, then at Mars, and finally at the aspects between these two.

Venus in the Signs

Let's start with Venus as it has a great deal to do with what you need in relationships. Venus is love itself. We all have a Venus, so we all need love, though the sort of love we all need isn't the same. Your Venus (according to its sign, house, and aspects) determines your basic romantic needs, the things you *must* have if you're to be happy in your relationships. Problems occur when these needs don't mesh with your want patterns, as is the case when you have Venus semi-sextile, square, quincunx, or opposition Mars. Although a relationship that fulfills your basic physical wants will be exciting and physically gratifying at the outset, as time goes on, those unfulfilled needs will become increasingly important. If not met, the relationship will almost always fail and disintegrate. Or, in the few cases where it does last, it will become a mere shadow of its glorious beginnings, a travesty of two people who share a house—and perhaps a bed—but nothing else.

Note that Venus is the feminine principle. It shows how you express (or in the case of blockages, repress) your feminine side. Regardless of your sex, you *do* have a feminine side; if you don't accept the feminine in yourself, you'll have trouble accepting the feminine in others, and chances are good that you'll have problems with women in general.

In a man's chart, Venus (and to a lesser extent, the Moon) shows his view of femininity and therefore helps determine what sort of woman he'll attract.

By looking at your Venus sign, you can get a basic picture of your romantic needs.

Venus in Aries
This indicates that you need a passionate lover and a very intense relationship. It may also be a sign that you don't express your feminine side very well, that you're more comfortable with steaming hot sex than tender loving care. For a woman, Venus in Aries was traditionally considered an impediment for many years because it was thought to be too pushy, too self-loving, and not sacrificing enough "for a woman." Put in more liberated terms, it merely means that you have a strong sense of self, which you value. You need a mate or lover who will respect this.

Venus in Taurus
This could mean a love of—and need for—possessions, financial stability, and/or physical comfort in a relationship. Certainly you need a steady income, so a seasonal worker, a struggling artist, or a mate who refuses to contribute to material well-being probably isn't going to lead to happiness ever after. On the other hand, you need—and expect—a lot of tender loving care, so a mate/lover who's totally caught up in career and money isn't a good bet either. You need someone you can admire, someone you see as strong, stable, and a good provider. Provider of what, of course, would depend on your upbringing and values (and possibly your sex, if you're over thirty).

Venus in Gemini
For you, happiness is learning and being on the move. You want to see it all, do it all. As a result, your needs may be different at different times and your affections can flit from person to person accordingly. What you need and value most are sincerity and good communication. Any relationship based on these things stands a good chance of surviving.

Venus in Cancer
You express your feminine side very well in the traditional sense. This means, if you're a

woman, you may be somewhat timid or passive, and are possibly looking for a traditional relationship. If you're a man, you may be the strong, silent type and again are probably looking for a traditional relationship. Venus in Cancer has a tendency to fear rejection and therefore tends to need a lot of security. Once love blooms, the relationship tends to be very intense; Venus in Cancer likes it that way.

Venus in Leo

This being a fire sign, we can expect Venus in Leo to need sexual attraction and passion. A very dramatic relationship is usually preferred; excitement is needed to keep the relationship out of a rut. A need to have children is often felt; if your mate can't stand kids this might become a serious sort of dissatisfaction. Venus in Leo very much needs tangible expressions of love—from "I love yous" to little gifts. Sometimes there's a tendency to feel, "If you love me, you'll prove it by waiting on me." Certainly your mate shouldn't be opposed to fussing over you a bit at regular intervals. One drawback to Venus in Leo is that you tend to have very high standards, to want it all. You need to feel your mate is special, because you need to feel *you* are special. Up to a point, this is fine, but if you set your standards impossibly high, you decrease the likelihood of fulfilling your needs.

Venus in Virgo

This is traditionally considered a cold—even frigid—Venus. It's not; but it doesn't have as great a need for flamboyant or passionate displays of affection as some of the other placements. What you do need is someone who can show his or her love in a variety of simple ways. You like simple but elegant candlelight dinners. You like a mate/lover who writes down or marks things he or she knows you'd be interested in when reading. You like a mate who feels details are important. You need to work, need to be busy. So the worst thing for you would be to be put on a pedestal and waited on hand and foot.

Venus in Libra

Because Libra is the sign of balance as well as one of Venus's signs, you should be able to very easily express either your masculine side or your feminine side, depending on what's being called for. You need marriage—or at the very least stable companionship—to be at your best. You also have a strong need for peace and harmony. Though you can "make do" with virtually anybody who's reasonably easy to get along with, in the long run you need someone who's doing something worthwhile with his or her life, someone whose experiences are worth sharing and meaningful in your eyes. This is because you can't love someone you don't also respect.

Venus in Scorpio

What you don't need in the romantic arena is a platonic relationship. You need an "all-or-nothing" sort of love, a mate/lover who puts you first. A strong sex drive is probable; certainly you value sexual attraction and compatibility more than some of the other placements. But this doesn't mean you'll settle for one-night stands or part-time love. Uh-uh. . . . If he or she wants you, he or she better be prepared to be there when *you* need him or her—not just the other way around! You also need to feel you know your mate's or lover's innermost secrets and, should he or she balk, you undoubtedly have several methods at your disposal for finding these out. So a partner who can't stand a certain amount of probing of his or her psyche is probably not for you.

Venus in Sagittarius

Although Sagittarius is a fire sign, I find it doesn't have the same sort of passion as Aries and Leo, and therefore you don't have as strong a need for sex. Rather, what you need is spontaneity, because your needs tend to change according to your circumstances. You also need a mate/lover who feels the chase is as important—or even more important—than the catch; you seem to need that game-playing, courtship stage more than most. For this reason, anyone who tries to turn things toward a serious commitment too soon is liable to wind up hearing "Good-bye" instead of "I do." You're as anxious for a stable relationship as the next one, but you want to be the one to stabilize it, and pressure to commit yourself only threatens your freedom and drives you away. Finally, you need a mate/lover who respects—and preferably shares—your philosophy of life.

Venus in Capricorn

This placement gets almost as much bad press as Venus in Virgo. Again, while slow to warm up, it's not devoid of passion; it just doesn't place sex quite so high on its list of priorities as do some of the other signs. You need a purpose in life, and the object of your affections must be able to fit himself or herself into this purpose. It's been said that Venus in Capricorn is inclined to marry for money or prestige. While this may have been true once upon a time, I've yet to see anyone with this placement who puts money or prestige at the top of his or her list of needs. But while you don't expect your mate/lover to necessarily improve your material or social standing, you won't allow him or her to interfere with it or cause it to deteriorate either. You need a mate who understands that an ivy-covered cottage teeming with kiddies just isn't enough for you, no matter how much love you get there. To feel worthy of love, you must feel you're doing something to deserve it. For you, that means proving yourself in the business world, or in your community, in some way by earning the respect of those beyond your four walls. But it also means you have a need to be sure of your mate/lover's respect and support. To work so hard only to have your partner undermine or belittle your efforts would be a crushing blow for you.

Venus in Aquarius

In traditional astrology, this could be keyworded as "'unusual romantic needs." These could lead to involvement with eccentrics of both positive (genius) and negative (drifter) types. It could lead to a need to remain single, or it could lead to a sudden commitment that lasts forever. On the other hand, the ruler of Aquarius being Uranus, the planet of divorce, there could be multiple marriages. The one thing any planet in Aquarius says is that anything (just about) is possible! Though it's hard to generalize on this placement, you do have some needs that must be fulfilled if you're to be happy. First of all, your mate/lover must be mentally stimulating. Secondly, he or she must allow you to keep the door to the future open. As soon as you feel trapped or locked in, romance stands a good chance of going out the window. You need to feel there's still more to uncover in your relationship, more truths to learn, more places to go together.

Venus in Pisces
You need a very emotional, sympathetic, affectionate mate, preferably one who's not too moody, as you may have a tendency to absorb negativity and let it get you down. You need love to be expressed constantly, consistently, and gently in order to be happy. Sex is probably less important than kisses and cuddles. If you have a belief in a specific religion or a branch of metaphysics or astrology or whatever, this needs to be respected as well. If these convictions are belittled or interfered with, you can become very mixed up and may feel you have to make a choice between your mate/lover and your beliefs. Sacrificing either for the other is apt to cause pain.

Mars in the Signs

Having taken a look at Venus to see what you need, our next step is to look at Mars to see what you want.

Mars in a woman's chart has traditionally been said to symbolize her ideal man. This is because Mars (and to a lesser extent, the Sun) shows your view of masculinity. But this isn't quite the same as a *concept* of the ideal man; rather, it's what you've been "programmed" to think men in general are like, and therefore shows how you expect to be treated by a man. You may, according to your conditioning and the way you've chosen to use your Mars, expect to be treated well or badly. In either case, you'll subconsciously motivate yourself to attract what you expect. Your sexuality, as symbolized by Mars, will then manifest accordingly in response to the men you meet.

In both sexes, Mars shows how you express your masculinity. It thus determines to some extent whether you fit the male or female stereotypes of your culture, including what sort of lover you are. Thus, a woman with a strong Mars in Aries might get labeled "castrating female"—not necessarily because she is a castrating female but because she's more active and more insistent than the traditional stereotype in terms of making her wants known and doing something about her sexual desires. Likewise, a man with Mars in Cancer may get labeled "wimp" because he prefers baking bread to going to singles bars and is more passive than our traditional stereotype in terms of going after what he wants. Neither of these placings is "bad" for sex or love, but society's biases may make it harder for these people to get what they want, particularly in the teenage years when peer pressure dictates conforming to the norm.

We've already looked at Mars by sign in a general way. Now let's see specifically what it says about your sexual drives and wants.

Mars in Aries

You want to be the dominant partner in the relationship and you want enough challenge to keep the relationship interesting. You want a courageous but not too headstrong mate/lover. You're a spontaneous lover, eager to be involved. If your love life isn't satisfactory to you, if relationships start well only to fizzle shortly thereafter, it could be you're coming on a bit too strong. Some men and women will tend to be overwhelmed by your ardor or even suspicious of it; they may mistake it for plain old lust or even desperation. You'd probably be happiest with a mate with Venus in a fire sign, Venus in the first house, or a Venus-Mars conjunction. A mate whose Venus conjoins your Ascendant might also be able to provide the kind of love you want.

Mars in Taurus

You want money. You also want physical demonstrations of love, and possibly tangible tokens of affection. You exude a strong sexuality, but it's low-key rather than overwhelming, tantalizing at the background of your personality. Though your wants at first glance may seem unromantic, you're not asking for something you're not willing to give. You may enjoy expressing your love by giving gifts, making gifts, or doing practical things for your loved ones. And you're also probably physical in your romantic expression. You undoubtedly like to give and get hugs, kisses, and cuddles in addition to sex itself. You'd be happiest with a mate who has Venus in an earth sign, Venus in the second house, or Venus sextile your Mars.

Mars in Gemini

You want conversation and you want your mate to be able to keep up with you mentally and physically. You're quite sensitive even though you often succeed at hiding it through witty statements or endless chatter about trivia. For this reason you sometimes sublimate your sexual energies into intellectual pursuits rather than run the risk of rejection on the romantic front. Contrary to popular belief, you're not uninterested in commitment; you're just afraid of being hurt. If there's good mental rapport between you and the object of your affections and if you understand one another's needs, all will be well. For best results, look for someone who has Venus in an air sign, Venus in the third house, or a Mercury-Venus conjunction. It also helps to have your Mars sextile or trine the other person's Sun, Moon, or Mercury.

Mars in Cancer

You want an emotional mate who likes to touch and be touched. Chances are good you also want a family. You can be a clinging vine and smother your lover if you're not careful. Your intentions are good—you want to protect, provide for, and take care of him or her so that your mate's every need will be fulfilled. You therefore do best with a mate who's a wee bit dependent, someone who wants a strong person who can advise and help when need be. In general, this rules out anyone whose overall chart shows a strong Aries (not at all dependent), Sagittarius (you'd cramp his or her style), or Aquarius (too independent for your tastes) influence. Choose someone who has Venus in a water sign, in the fourth house, or conjunct his or her Moon for best results. You mate's Venus on your Nadir might also be a good choice if you want a comfortable, relaxed, stay-at-home sort of relationship.

Mars in Leo

You want—you demand—to be the center of attention. You also want to achieve something in life and won't take kindly to anyone interfering with your goals. Your approach to sexuality is fairly direct. You don't necessarily come on strong, but you make your wants known and aren't much on playing games. Probably you're faithful; certainly you expect fidelity from your mate or lover. For best results, choose someone who has Venus in a fire sign, Venus in the fifth house, or a Sun-Venus conjunction. Try to see that this Venus isn't under stress from natal Uranus or from your own Uranus.

Mars in Virgo

You want a healthy mate/lover. While not cold, you probably don't want a terribly sexually-demanding partner because you feel you have to have energy for other pursuits as well. You're very careful about whom you get involved with. You're also careful where and how you get involved; you're nothing if not discreet! This doesn't necessarily mean you lack spontaneity; it does, however, mean that you're not one to rush into things. Impulsiveness tends to leave you feeling guilty; lack of preparedness worries you. So when you become involved, you invariably know exactly what you're doing. And while you may like variety, you don't like one-night stands. Being discriminating is your style, and it will work for you as long as you don't take it to extremes. You'd be happiest with a mate or lover who has Mars in an earth sign, Mars in the sixth house, a Mercury-Mars conjunction, or a Mercury-Mars sextile.

Mars in Libra

You want companionship. You want to avoid loneliness and will pay just about any price (in terms of compromising or altering your needs) to do so. However, you also want a growth-oriented relationship, so you're apt to be quick to instigate change if you feel things are getting too humdrum. You don't mind taking the initiative to get a new relationship off the ground; you're a little less quick to initiate a break-up unless there's either someone else waiting in the wings or your lover has refused to budge and things just can't be improved enough to restore harmony. You'd do best with someone who has Venus in an air sign, in the seventh house, or trine your Mars.

Mars in Scorpio

You want sex. You also want to be allowed to be a little or a lot self-indulgent on occasion. But there's a paradox here. Even though you want sex, and actively seek it, because Scorpio is a passive sign, you're not entirely comfortable proclaiming your wants and openly pushing for them. In fact, you may try to deny them even to yourself. From what I've seen, Mars in Scorpio, particularly in women, is inclined to stifle or downplay its sexual energies for lengthy periods. Sometimes this is due to an overly moralistic upbringing, sometimes it's because your sexual energies are sublimated into psychology or the occult, and sometimes it's merely because of a lack of a suitable sexual partner. Anyway, because of this tendency, when your needs do make themselves felt, you may go through long periods when you're almost insatiable. You do best with someone who can adjust to both your "on" and your "off" periods. Choose someone who has Venus in a water sign, in the eighth house, or conjunct Pluto for best results.

Mars in Sagittarius

It's hard to say what you want except that you want what you believe in. And you'll fight to get it. "It" might be foreign travel, higher education, specialized knowledge, or merely freedom to come and go as you please with no strings. You tend to approach virtually any involvement in a wide-open, anything-goes fashion. For this reason, a lot depends on your philosophy of life. There can be beautifully spontaneous relationships that deepen without losing their zest as time goes on, or there can merely be a series of rather superficial relationships—fun while they last, but not lasting long. Your best chances of happiness would come from involvement with someone who has Venus in a fire sign, Venus in the ninth house, or a Venus-Jupiter conjunction. People with fire or air predominate in their charts will probably be more compatible with you than those who have an emphasis on earth or water.

Mars in Capricorn

You want advancement and recognition. You're inclined to make quite specific demands on your mate or lover, these demands having to do with the sort of support you need to advance in life and gain the recognition you want. But you're not all take and no give—you almost certainly demand a great deal of yourself when it comes to playing romantic roles. You tend to want the sort of love life described in novels, complete with the early phase of fireworks alternating with excruciatingly painful doubts, the eventual vows of eternal love, and a happy-ever-after ending. But this doesn't always happen, and even if it comes close, there'll undoubtedly be some humdrum or even turned-off periods. Expect these—and don't take them too hard; they happen to everybody. You'd do best with someone who has Venus in an earth sign, Venus in the tenth house, or a Venus-Saturn conjunction. (Note: Contrary to popular belief, Venus conjunct Saturn doesn't automatically imply frigidity. In fact, I find it more often gives romantic intensity. While people with this aspect may be a bit slow to warm up, when they do there's generally plenty of romantic ardor!)

Mars in Aquarius

You want friends and may be more comfortable with a platonic relationship than a passionate one. You want an interesting mate or lover and a relationship that evolves and renews itself continually. Being independent, you don't want a clinging vine or a terribly possessive partner. Your style of romantic/sexual expression is multifaceted, making it hard to generalize about you. Basically, you're idealistic; you may therefore do a lot of "market research" or keep your options open longer than some of your peers. You'd do best with someone who has Venus in an air sign, Venus in the eleventh house, or a Venus-Uranus conjunction or sextile.

Mars in Pisces

You want time for contemplation. You don't want a terribly aggressive or sexually demanding mate or lover, and your sexuality and type of romantic self-expression is subtle. You're not aggressive yourself and don't like people who come on strong. Your overtures to possible romantic partners are sometimes so low-key that they get misinterpreted—or even missed altogether. They may, in fact, be more instinctive than conscious much of the time, making it hard for you to figure out what's happening yourself. For this reason, you need a very perceptive and rather sensitive kind of lover, someone who's willing to spend time and energy drawing you out of yourself. Someone with Venus in a water sign, Venus in the twelfth house, or Venus conjunct Neptune would be ideal.

Armed with these very basic insights into your sexual-romantic wants, needs, and style, you should be able to get some inkling of why you attract the sort of lovers you attract. But before looking at Venus-Mars aspects, it might be a good idea to clarify a couple of things.

First of all, no sign is in itself bad for love. What's bad is to be out of touch with your wants and needs, or to be aware of them but repress them in an attempt to be what you think others expect you to be. Granted, some Venus-Mars combinations are trickier to work with than others. For example, if you have Venus in Aries, you may need an outlet for your sexual passions. But if you have Mars in Pisces as well, you may be motivated strongly to repress this need. Maybe your religious background left you feeling sex is wrong. Or, if you're a woman, you might have bought that old cliche about nice young ladies never expressing an interest in sex—at least not before marriage. Or maybe you've bought one of those old lines about sex robbing you of your creative or psychic powers. These attitudes may seem quaint in today's supposedly liberated age, but they still exist and one or more of these influenced your upbringing.

Sex isn't the only need that can be repressed. What if you have Venus in Taurus and Mars in Aquarius? Venus in Taurus needs money and security, but Mars in Aquarius wants to believe it can detach itself from materialism. Or what if you have Venus in Gemini and Mars in Cancer? Venus in Gemini needs to learn about love and life by trying a variety of things, but Mars in Cancer wants security, and what's secure is generally what's familiar.

How do you reconcile conflicting wants and needs? How do you tell which will predominate? Well, if you have more aspects to Venus than to Mars, chances are you'll take a more passive approach to fulfillment. Harmony and stability are high on your list of priorities, so you're apt to be more willing to try to be what your environment seems to expect you to be—even if that's not in keeping with your wants or needs. If, on the other hand, you have more aspects to Mars than to Venus, you'll take a more active approach to fulfillment. You'll tend to put more emphasis on your own wants and needs and will be more courageous about expressing them—even if it means at times upsetting those around you.

If Venus and Mars are both in feminine signs, you'll find that a passive, wait-and-don't-upset-the-applecart approach works best in terms of bringing you what you want. Impulsiveness will throw off your timing and threaten your credibility; fighting or demanding will probably get you labeled as spoiled. If, on the other hand, Venus and Mars are both in masculine signs, you'll do best by taking action to get what you want, and firmly leading people in the direction you want them to go. Hesitate, and others will say you're lost or think you've got no backbone. Give in to keep the peace and you'll be treated like a doormat. If you have Venus in a masculine sign and Mars in a feminine sign or vice versa, you can swing back and forth as you feel circumstances warrant without getting too much flak from anybody. You have a choice.

Problems arise when society intervenes to block your most natural mode of expression. If, for example, you're a female born before 1960, chances are you were expected and conditioned to act in a "'more-aspects-to-Venus, Venus-and-Mars feminine" way. This is fine if this is your natural mode of expression. If it isn't, pressure may be put on you to repress the masculine energies in your chart. When this happens, they become perverted; instead of being seen as perfectly natural and healthy assets, they become traits to be feared. Having nowhere to go, they remain hidden until they can express themselves acceptably. In this case, acceptable expression is apt to be construed as projecting them onto an available and willing man. But then yet another problem rears its head. Having been conditioned for so long to believe these qualities were bad, how can you possibly accept them in another when you couldn't accept them in yourself? No matter how attractive they are, as long as you give them a negative connotation, you're going to have an element of negativity in any relationship you might have.

So if relationships are a problem, it may not be the other person—or even the aspects—at fault; it may be merely a case of rejecting a part of you and in doing so giving your subconscious—and other people—the message that you're not worthy of their love.

Venus-Mars Aspects

Accepting the necessity of both the masculine and the feminine functions of your personality—of your need for passion, risk, challenge, and your need for tenderness, reassurance, and nurturing—is essential. You may need more of one than the other, but you need at least some of both. This said, the next step is to look at the Venus-Mars aspects in your chart in order to see how they affect the evolution of your relationships.

Relationship Venus Conjunct Mars
You have very strong emotions. Because of their intensity, you find it hard to treat relationships lightly; everything, even the smallest gesture, has importance to you. This is true in almost all

relationships, and especially so when it comes to sex and romance. If there are problems in your love life, it could be because you're being too pushy or coming on too strong.

This aspect stimulates affections in a way that causes you to want an all-consuming, "forever" type of relationship, but oddly enough, it also gives a flirtatious streak that continues even after you've entered into a satisfying relationship. I think that people with this aspect feel it's especially important to remain desirable—not just to a mate or lover, but to the opposite sex in general.

As the conjunction is a uniting aspect, and Venus and Mars are the female and male principles, respectively, you might think, as I once did, that it's a portent of a lasting marriage. However, I've learned this isn't the case. While people with this aspect want a lasting relationship very badly, they suffer their fair share of divorces and more than their fair share of squabbles and separations. There's a difference between this aspect and the square and opposition, though, in that there seems to be greater reluctance to make the first break. Jealousy—sometimes justified, sometimes not—on one side or another seems to be the primary cause of discord.

If you have Venus and Mars in the same sign but not conjunct, the foregoing may apply to you. However, you'd be less pushy.

Relationship Venus sextile Mars
You probably get along well with the opposite sex. They like you, and you like them, with one exception—you tend to find passive people quite frustrating. Your approach to love is one in which you throw yourself into each new romantic experience. You don't like to do things halfway, and once involved, you put everything you've got into making a relationship work.

Occasionally, sexual-romantic energies are sublimated into some sort of artistic activity. This is most common when air is the predominant element in the chart.

Normally, however, you're a loving person who wants and needs a sexual outlet. For this reason, a less than satisfying sexual relationship could lead to infidelity.

If you have Venus sextile Mars by sign but not by aspect, the foregoing may apply, although your need for a sexual outlet would be less strong.

Relationship Venus Square Mars
Your sexual-romantic feelings are very intense. In fact, you have to be careful that you run them, instead of the other way around.

This is a disruptive aspect that works primarily in the love life area, although it can have repercussions in financial and other areas. It's one of the classic indicators of a propensity for divorce and romantic-marital difficulties. It also occasionally coincides with the death of a mate during marriage, but this is less frequent and there would have to be substantial confirmation of this tendency found elsewhere in the chart before anything of this sort should be expected.

People with this aspect are demonstrative, warm, and generally faithful when committed, but they demand a lot in return. In extreme cases, when needs aren't met, dissipation of various types can result, with infidelity, gambling, and/or excessive drinking being the most common form of misdirection of romantic-sexual energies.

If you have Venus square Mars by sign but not by aspect, some of the foregoing may apply to you. However, you'd be less inclined toward dissipation and slightly more inclined to find a constructive way to meet your romantic-sexual needs.

Relationship Venus Trine Mars

You have an enthusiasm for romance and sex that other people find attractive. You like to be with the opposite sex and show it. They, in turn, like you for being yourself and not playing games with them.

Your relationships should be good, as should your sex life. Not that this aspect is a guarantee of romantic success; there can still be an unhappy love affair or two but it's certainly a plus and can help offset other stressful romantic aspects you might have.

One influence with this aspect that could need watching is the fact that you tend to be susceptible to flattery. While you don't play games with other people's affections, you're a bit vulnerable to being caught up in other people's games simply because you mean what you say and therefore assume others also mean what they say. If there are problems in your love life, it could be a sign that a little more discrimination is needed.

If you have Venus trine Mars by sign but not by aspect, the foregoing would apply to you to a lesser degree. You might not be quite as averse to playing games as those with the trine aspect, though.

Relationship Venus Quincunx Mars

Your emotional self-control is erratic. You find it difficult to react objectively to the opposite sex because your emotions get in the way. Even after getting involved, your feelings about your loved one are often mixed. There's a "what am I doing with him or her anyway?" feeling that comes and goes even when the relationship is running relatively smoothly. So your romantic and sexual relationships tend to be a bit stormier than those described under the conjunction, though not generally as stormy as they might be if you had the square or opposition.

You're inclined to make snap judgments when it comes to members of the opposite sex. You want what you want as soon as possible—yesterday would be ideal! There's also a tendency to want to have your cake and eat it too.

Although you're a passionate, loving person, you often find some facet of your love life disappointing. This is invariably because you seeking perfection. Be realistic about your loved one's virtues as well as his or her flaws and things should go smoothly as long as you're consistent with your expectations.

If you have Venus quincunx Mars by sign but not by aspect, some of the foregoing may apply to you, although you'd be more inclined to wait for what you want instead of demanding it quickly.

Relationship Venus Opposition Mars

You have an intense emotional nature. Often your feelings about the opposite sex are ambivalent. In love, there's often a tinge of "can't live with him or her; can't live without him or her." You love passionately and hate passionately—sometimes the same person! There's a great deal of impatience in relationships, especially romantic relationships. This sometimes stems from a lack of trust in the opposite sex.

Often this aspect signifies a person whose childhood was unhappy. This may have been a result of difficulties with the opposite-sex parent, a strife-filled or broken home, or being ridiculed or ignored by the opposite sex in the teenage years. Occasionally it results from a too-early sexual experience that came before the person was mature enough to understand and handle it. Anyway, there's often a carry-over of these childhood experiences and the emotions stemming from them into the adult years. Often therapy can banish these "ghost emotions" and lead to a more satisfactory love life.

Sometimes sexual-romantic energies are sublimated into artistic activities. For those who are inclined to be excessively intense, involvement in art or music can bring strong emotions down to a manageable level.

If you have Venus opposition Mars by sign but not by aspect, the foregoing may apply to you. However, less energy would be expended in passionate hating; you'd be more inclined to just break up and go on.

Relationship Venus Semi-sextile Mars

This is a vague aspect that seems to contribute to nontraditional sexual-romantic attitudes. It should be considered as a reinforcer of other similar aspects rather than a tendency in itself. Semi-sextiles by sign (out-of-orb) don't seem to have any relevance at all.

Venus and Mars by House

It may also be useful to look at Venus and Mars by house to see what you need and what turns you on.

First House

Venus here says you need to lead in some way. You also need to feel you're attractive.

Mars here says being on the move turns you on. Being kept waiting for an answer turns you off.

Second House

Venus here says you need possessions and may place a particularly high value on good looks in a potential partner.

Mars here says you want stability and a mate/lover who wants to be possessed or wants to defer to you.

Third House

Venus here says you need the sweet nothings, the cards, the love notes that are a part of courtship. Talking about love and talking about your relationship is important to you.

Mars here might say you want a well-read lover, but probably it's also saying you want a sexually experienced lover. You probably have quite definite opinions about what you do and don't enjoy and quite specific requirements about your mate/lover's attitudes, sense of humor, interests, and so on. More about these can be told by looking at your Mars sign and aspects.

Fourth House

Venus here says you need a house. You also need a mate with whom you can relax and let your hair down. This seems to be a placement that's actually relieved when the honeymoon's over because you don't feel you can fully know a person until you've lived and shared together for some time.

Mars here says your wants are pretty basic—love, loyalty, a home, an even temper. One thing you want is a sense of shared past or similar background. You may insist on knowing your mate/lover's romantic history before committing yourself.

Fifth House

With Venus here you need loyalty, enjoyment, and creativity in your relationship.

Mars here says you want to be entertained. You therefore may have a greater-than-otherwise desire for sexual pleasure. But more than this you want a mate with a zest for living.

Sixth House

Venus here says you want to serve, help, or wait on your mate/lover because you need to be needed. On the other hand, you don't want to lose your health or your physical shape in the process.

Mars here says you have quite specific requirements; notably you want a mate who is discriminating, enjoys going into the details and is sexually cautious without being hung-up and inhibited.

Seventh House

You need a mate/lover who finds you attractive. Your own happiness tends to be very dependent on your partner's happiness. So what it comes down to is you need to please if Venus is here.

If Mars is here, you want a monogamous relationship and you want a mate/lover who can be depended on to be there when you need him or her. But at the same time, you need a social life, so woe be unto the mate/lover who tries to make it just the two of you forevermore.

Eighth House

With Venus here, you need a very intense relationship, one that you can immerse yourself in totally. You need to feel you have a relationship that's worth working at. You also need a mate who has an understanding of human sexuality.

With Mars here, you want sexual activity, Without it you become very tense and start to feel empty or lethargic. But casual sex isn't apt to be completely satisfying; you want understanding as well.

Ninth House

With Venus here, you need someone who's different from you and can help you broaden your outlook. You might find travel, politics, or a Ph.D very exciting and attractive. Whatever it is, the key is you need someone who's doing something you feel is interestingly exotic.

With Mars here, you might want a mate/lover who enjoys nature and the outdoors. Or you might want a partner who's a connoisseur of culture or has some sort of highly specialized knowledge. Or you might want someone with whom you can play the role of sexual mentor.

Tenth House

With Venus here, you need a respectable—if not prestigious—social standing. You also need some sort of distinct line drawn between your private life and your public life, because, although you like attention, life in a goldfish bowl—if constant—tends to pall after a time.

With Mars here, you want help planning or scheduling your life. You want—though you may not realize it—a mate who will push you to take regular breaks from your public life and career, one who will encourage you to express yourself rather than be a mouthpiece for, or puppet of, some segment of your environment.

Eleventh House

You need to help others—need to be a helpmate—if you have Venus here.

With Mars here, you want to be a humanitarian. You may also want to be considered an accomplished lover.

Twelfth House

With Venus here, you need music, candlelight, and all the other little niceties that put you in the mood. But you may also need some sort of bittersweet element, some sort of problem to surmount or sacrifice to be made. Negatively, this placement can symbolize an unhealthy need for suffering stemming from early-life (some even say past-life) conditioning.

With Mars here, you want a certain amount of time to yourself. This placement tends to give a fear of rejection; therefore, you want to take things slowly and carefully.

Conflicting Needs and Wants

The only missing piece of the Venus/Mars puzzle is what to do about conflicting wants and needs. Alas, while there are pointers, the answer to this one won't be found in any book. It lies within you and within your chart. The only way you'll find it is by taking a long hard look at yourself, your strengths, your weaknesses, your wants, and your needs, accepting these, and in the case of conflict, prioritizing them. In prioritizing, try to put needs before wants. Wants may, when they manifest, be more insistent on fulfillment, but wants will tend to come and go. Needs, on the other hand, last a lifetime. Maybe you can't get everything you want, but you must, to be happy, get what you need.

This information is all very well when you're sitting with your romantic interest's chart in front of you. But what about those times when you're out and just happen to meet someone utterly fascinating? Is there any sort of quickie way to decide whether or not a relationship is worth pursuing?

The answer is a qualified yes. As you know, to get a true picture of a person, we must look at his or her whole chart, so in that sense, an instant answer just isn't possible. But if you're of two minds about someone, there is a technique that can be used to decide whether or not to pursue the relationship—at least long enough to feel comfortable about asking for the person's complete data. This information is given in Appendix II.

Chapter 3

Occupational Motivation

When I first wrote *The Mars Book*, Mars was not generally considered to be an important career significator unless it was in one of the career houses (houses two, six, or ten), ruling one of the career houses, or aspecting the Midheaven. However, that was about to change, thanks to the work of Michel and Francoise Gauquelin, French researchers who had actually set out to disprove astrology. The Gauquelins collected approximately 16,000 birth certificates of eminent European professionals in ten occupational categories. They then set out to see if there were any correlations between planetary positions and professions in these charts. For some planets, such as the Sun and Mercury, there were no statistically-relevant correlations. For others, including Mars, there were. However, there was something weird about these correlations. Mars, and the other planets for which there were correlations, were generally not found in the angular houses, as expected. Rather, they were frequently found in the cadent houses. The significant chart sectors were named Gauquelin sectors. There are four of them—two primary sectors and two secondary ones. The primary sectors consist of the first third of the first house (roughly ten degrees) plus thirty degrees behind the Ascendant into the twelfth house, and the first third of the tenth house plus roughly thirty degrees behind the Midheaven into the ninth house. The secondary sectors consist of roughly ten degrees (one third) of the third and sixth houses just before the Descendant and IC.

Suffice it to say that neither the scientists nor the astrologers were particularly enchanted with these findings. Some astrologers dismissed them outright because they flew in the face of tradition. And the scientists, who for the most part had no inclination to accept anything that might validate astrology, cried foul, fraud, and falsification. And at the center of this controversy was Mars. Gauquelin found Mars to be in one of these Gauquelin sectors significantly beyond the chance level in the charts of athletes, soldiers, physicians, and business executives. Three of these occupations are among those that are typically associated with Mars. While there were other findings for other planets, it was the Mars findings that were most striking. And since this book is about Mars, I will refer you to *Cosmic Influences on Human Behavior* by Michel Gauquelin, *Psychology of the Planets* by Francoise Gauquelin, and *The Tenacious Mars Effect* by Suitbert Ertel and Kenneth Irving for further details on the Gauquelins' research and their findings in terms of the other planets. Meanwhile, be aware that if you have Mars in one of the sectors just described, then your Mars is apt to be stronger than usual and shows some aptitude for the careers mentioned above.

While not everyone has Mars in a Gauquelin sector, I feel that no matter where Mars is in your chart, it will have a great deal of bearing on your career. Why? Because as you've already seen, Mars has a great deal to do with your energy level and your motivation in general—and both these factors in turn have a great deal to do with success or failure in any profession you might choose. So let's look at Mars as a career indicator, first of all generally, and then according to its sign and its relationship with Saturn, another biggie on the career front and another planet for which Gauquelin found significant correlations.

Career Mars in the Signs

Some people are more comfortable leading the parade, making things happen, and putting their energy out there. Generally people of this type have Mars in fire or air signs. When your Mars is in fire or air, your tendency is to use an assertive approach in dealing with others. You're generally happier in a leadership capacity or working on your own at your own pace than you are as a rank-and-file worker. You don't want to be waiting for orders; you want to be giving them. You generally go out and get what you want. If, on the other hand, your Mars is in earth or water, you may be just as happy to let someone else do all the demanding and order-giving and simply collect your paycheck on payday. Support or service positions may appeal to you just fine, and you may prefer the status quo to initiating changes or pushing new projects.

Fire sign Mars people tend to be optimistic and are generally good at getting others to do what they want. Earth sign Mars people are practical and generally have good reasoning skills. They therefore are better than average problem-solvers. Air sign Mars people tend to be studious. They're good at stimulating others to think and tend to thrive on report-making and the like. Water sign Mars people tend to be sensitive and understanding—sometimes too understanding for the company's good in the long run.

Cardinal-sign Mars people tend to work quickly and are good at starting things (though Mars in Cancer sometimes needs some encouragement at the outset and is thus a bit slower on the uptake). They generally work well on their own, although Mars in Libra would really be happier (though not necessarily more productive) with a partner. Fixed-sign Mars people like to see projects through from the beginning—or almost the beginning—to the end. They tend to be exceedingly uncomfortable walking in on the middle of a project and feel a need to go back to the start and run through from the beginning before proceeding further. And if you take them off a job before it's done, you may have to hire a security guard to keep them from going back and trying to finish it. They won't last long in a physically uncomfortable work environment; extremes of heat or cold, noise, standing on hard floors all day, et cetera, will generally send them looking for greener pastures. Mutable-sign Mars people are at their best when versatility is called for. They're also the best talkers, so they make good wheeler-dealers of all types.

Career Mars in Aries

You have initiative and a high energy level, although to be at your best you must enjoy what you're doing. You also have executive ability, although at times you can be a trifle domineering, which can hold you back. You fall down on repetitive jobs—when the challenge goes, so does your motivation! Some good career choices for you would be: construction, mechanical engineering, highway planning, sheep farming, wool processing, politics, or psychology.

Career Mars in Taurus

You aren't too well-suited to business partnerships, partly because of a self-indulgent streak that can cause resentment and, ultimately, friction, and partly because of your stubborn streak. So you're better off working for yourself or working somewhere where you're allowed to do pretty much as you please. You're a hard worker as long as you see rewards on the horizon, but a life of all work and no play is definitely not a part of your game plan—you like the occasional long lunch, a decent vacation, and a good salary with which to get a fair number of life's goodies. With proper training, the following career areas could meet your needs: finance, banking, art, music, architecture. building, cattle breeding, farming, routine office work (if the pay is good and there's potential advancement).

Career Mars in Gemini

Your energy is primarily mental; your physical energy level is erratic. You're not always practical, and for this reason you need a trusted advisor or second-in-command to help you make the most of your potential. Some career areas that offer promise are: transportation industry, travel agencies, book buyer or seller, journalism, publishing, printing, advertising, communication, teaching, clerical work, and graphology (handwriting analysis).

Career Mars in Cancer

You're able to succeed in a number of career areas, but would be best off to aim for a position with a title, preferably a position that involves serving the public in some way. You're motivated to gain prominence and aren't too well-suited to partnership unless your partner is a silent one. You are, however, a prime candidate to be owner-manager of your own business. You're not suited to assembly-line work in a factory; nor are you suited to office work where you're one of several doing the same sort of work. You need to express your individuality in the career area. Some good career areas for you are: catering, cooking, bartending, advertising, office work (but not the very junior positions unless you really feel needed and there's scope for advancement once you prove yourself), shopkeeping, plumbing, nursing, hotel work, and fishing.

Career Mars in Leo
You have what it takes to work for yourself if you want to. Your energy level is high and your productivity generally good even though you're not the fastest of workers. A tendency toward arrogance needs watching if you employ or supervise others (or hope to eventually). Otherwise, you "sell yourself" well and generally work to make the most positive use of your potential. Some career areas to consider are: art, acting, stock exchange work, paper-making or selling, jewelry-making or selling, cosmetic sales or make-up artist, or the dance.

Career Mars in Virgo
Working for yourself or working where you have the freedom to do your own thing would be the best, although you're not averse to working your way up the ranks to a middle-management position if necessary. You're probably a hard worker and a meticulous one, although you may at times be less than honest if your job, a promotion, or a big commission is on the line. You're not at your best as "the big boss" as you're inclined to be rather critical of your staff yet at the same time are loathe to fire anyone. Some good career areas for you would be: the food industry, food inspector, clerical work (preferably a '"person Friday" in a small office), law enforcement, editing, teaching, chemistry, photography, and nursing.

Career Mars in Libra
You need to use discrimination in choosing a job if you don't want to experience disappointment. Anything involving hard physical labor is out unless the end result is a work of beauty. Secondly, you don't take orders very well, so menial work isn't a good bet unless it's of a temporary nature. What is good for you? Try the following areas: clock-making or repair, artistic work, politics, sales, brokerage work, modeling, hair stylist or barber.

Career Mars in Scorpio
You can do well on your own. You're also a good company person, provided you're treated well. You do especially well in service industries. You don't do so well in partnerships because you can be stubborn and dictatorial, neither of which will help you when there's a stalemate. Some good career areas for you are: medicine, surgery, engineering of any type, ice-making, refrigeration, law enforcement, banking, mining, undertaking, or brewing.

Career Mars in Sagittarius
You're best off working for someone else in a salaried job. You have a high energy level and are a self-starter, but meeting deadlines isn't your strong suit. You have a tendency to get sidetracked and spread yourself too thin. In a corporation, this tendency will be squashed pretty quickly and therefore will be kept under control by you if you value your job. But when self-em-

ployed or working freelance, with no one to make you account for your time, your unreliability can lead to more than one setback. Curb that tendency to scatter your energies, and success will be yours! With proper training, you could do well in the following areas: legal work, religious work, publishing, diplomatic work, publicity, communication, lecturing, philosophy, travel, exploring, or flight engineering.

Career Mars in Capricorn
You're a hard worker, having both integrity and perseverance. You're probably very ambitious and a good organizer too. So your ability to run or supervise a business is no doubt very good. Best career areas for you are: office work where there's scope for advancement, mining, government work, politics, mathematics, or osteopathy.

Career Mars in Aquarius
You need a profession rather than a job. It should be predominantly mental work rather than physical. Since you have a rather strong individualistic streak, general office work is not a good choice. Your dislike of the nine-to-five routine is apt to show in this type of setting and make you unpopular. You can successfully work for yourself, though, or in an area where you're given a lot of freedom. Some potentially good career areas for you are: astronomy, gymnastics, art, electricity, lecturing, and working with formulas in general (as in science and astrology).

Career Mars in Pisces
There are quite a few directions you can go in order to be successful, assuming you're willing to get proper training. Normally you do best working under others in a public service capacity, though. No matter what you do, you're apt to experience some dissatisfaction because with Mars in Pisces there's a distinct tendency to prefer dreaming to doing. Some good career areas for you are: automotive engineering, diving or diving instruction, textile industry, nursing, medicine, prison work, psychiatric work, movie work, the shoe trade, work connected with the manufacture or sale of alcohol, or chemical engineering.

These lists are of necessity brief. They're meant to get you thinking about what you can successfully do, not to restrict you to hard and fast choices.

Mars and Saturn Aspects

Next we'll look at aspects between Mars and Saturn, since this combination speaks of your motivation (Mars) to rise to the top in your career (Saturn).

Career Mars Conjunct Saturn
You often feel frustrated. If your frustration builds up for too long, you tend to become reckless. Yet at other times you're capable of showing a great deal of self-discipline. Used properly, this

aspect regulates your energies; you're self-reliant, patient, and very practical in your actions. Misused, your frustration can make you vindictive when thwarted.

You tend to experience less frustration working with things than you do working with people, particularly if this conjunction is cadent (in the third, sixth, ninth, or twelfth house). Some good career choices for you would be: lithography, boilermaker, sewage treatment work, repairing business machines, repairing industrial machines, foundry work, stenography, building custodian, carpenter, or aircraft mechanic.

If you have Mars and Saturn in the same sign but not conjunct, the foregoing may apply to you, but you'll be less self-reliant and your periods of frustration less frequent.

Career Mars Semi-sextile Saturn

You're slightly materialistic and have probably had some struggles connected with the involved houses. This aspect has a mildly detrimental effect on flexibility. It doesn't signify any particular career aptitudes; nor does it have much impact if out of orb.

Career Mars Sextile Saturn

You're self-reliant, but you may have trouble dealing with your anger. This can be a real problem unless somewhere along the line you've learned to release your anger in a positive way by sitting down and discussing whatever caused it. This aspect implies stability of action and steady application of action. However, when you're angry, you tend to work in fits and starts while seething inside. This is why repressed anger is so dangerous for you.

You work well with ideas. And as this is the most adaptable of the Mars-Saturn combinations, there are many career options open to you. Some examples are: city manager, market researcher, teacher's aide, travel agent, mining engineer, welder, statistical clerk, health inspector, kindergarten teacher, and air traffic controller.

If you have Mars sextile Saturn by sign but not by aspect, the foregoing may apply to you. However, you'd be slightly less adaptable.

Career Mars square Saturn

This is an indication of a lot of hard work in life. There are apt to be restrictions of some sort connected with the involved houses. On the plus side, you're courageous and assertive. On the minus side, you can at times be too aggressive for your own good and may lack sympathy. Jobs that require a consistently bright, bubbly personality, or a constant involvement with complaining people, or people who are chronically down on their luck are contraindicated, particularly if your chart is earth-predominant or if this aspect involves the first decanate of the cardinal signs.

It helps if you can see tangible results to your work, particularly if your chart is earth-predominant, or if your Mars-Saturn square is in fixed signs, or if it's a part of a T-square or Grand Cross involving Mercury. Some possible career choices for you are: boilermaker, mining engineer, occupational therapist, speech pathologist (for these two, at least three factors—planets, Ascendant, and/or Midheaven—should be in water signs), foundry work, machinist, stenographer, barber/hair stylist, carpenter, and railroad track worker.

If you have Mars square Saturn by sign but not by aspect, some of the foregoing may apply to you, but you'll be more sympathetic.

Career Mars Trine Saturn

You may not be an especially imaginative or original worker, but you're loyal, determined, and persistent; in other words, a potentially good company person. You're a hard worker and expect to be rewarded for your efforts, preferably with cold, hard cash. You prefer not to take chances. You believe that success stems from hard work, not luck! This is fine as long as you don't let your cautious streak cause you to turn down potentially valuable opportunities for increased income or status. Not everything can be a sure thing, you know!

You want to organize your job so that you have as much control over your working life as possible. So while you may not be an idea person per se, you do need an opportunity to express your feelings about your job and your working environment. This is particularly true when Mars is cadent. Some good career choices for you would be: mining engineer, occupational therapist, speech pathologist, actuary, advertising work, elementary school teacher, high school teacher, soil conservation work, aerospace engineer, or work involving electronic switching systems (for example, in the telephone industry).

One problem that sometimes crops up with this aspect is a tendency to withhold information to protect your own interests or because you assume—rightly or wrongly—that you're "sparing" other people the need to have to face something potentially unpleasant. This is dishonest. If you feel this could be a trait of yours, curb it now. Otherwise, it's liable to get you into hot water.

If you have Mars trine Saturn by sign but not by aspect, the foregoing applies to you to some extent, although you have less willpower than those with the trine by aspect.

Career Mars Quincunx Saturn

You can have difficulty in the career area due to a conflict between your responsibilities and your wants. For this reason, you may sometimes feel that people are purposely trying to frustrate you. Generally they aren't, but the responsibility you feel toward them makes you angry and causes you to feel if it wasn't for them, you could be out doing what you want to do. Your anger stems not from their dependency on you, but from your own ambivalence about your priorities. In other words, you get mad at yourself for being indecisive and then blame others for making you have to decide. You project your anger with yourself toward others. This is not an aspect that gives discipline. It contraindicates military and civil service careers as well as many types of mathematically-oriented careers.

People with this aspect tend to project confidence and may be quite self-assured as long as they don't have to try too many new things. Fear of the unknown sometimes leads to various forms of intolerance, which of course can limit career progress. Intolerance would also narrow your options in life, effectively limiting your need to make decisions.

Unless your chart is mutable-predominant, you're probably best off avoiding occupations where you have to work as part of a team. You do well working alone, though—or for yourself—in tasks requiring initiative. Some examples are: sewage treatment work, travel agent, business machine repair, industrial machine repair, computer servicing, machinist, stenographer, building custodian, elementary school teacher, airline ticket agent, librarian, architect, and geologist.

If you have Mars quincunx your Saturn by sign but not by aspect, the foregoing may apply to you to some extent. However, you'd be more tolerant and perhaps more aware of where your frustration really comes from.

Career Mars Opposition Saturn

You're hard-working, but tend to take an all-or-nothing approach to things that can ultimately be destructive to your goals. Sometimes this aspect signals workaholic tendencies; other times, it indicates refusal to follow orders. You often have the feeling that people are trying to stop you from doing what you want to do—or even from being yourself. You resent this, but instead of talking about your feelings when they occur, you tend to hold your emotions in for as long as possible. As a result, you often blow up over trivial things, which of course doesn't earn you any appreciation. You've probably been held back by parents or other authority figures more than once. These events may well have caused bitterness or dislike of authority figures in general.

Unless your chart is mutable-predominant or has most of the planets on the Descendant side, you tend to be uncomfortable working under close supervision and do not like having to go through formal job performance evaluations. You're best off directing other people's activities, perhaps in one of the following capacities: city manager, teacher's aide (if you're actually supervising or helping children, but not if you're merely a titled go-fer who cleans up messes), mining engineer, occupational therapist, speech pathologist, industrial or office supervisor, bank officer, hotel housekeeper, elementary school teacher, air traffic controller, and aerospace engineer.

If you have Mars opposing Saturn by sign but not by aspect, the foregoing may apply to you, but there is less of the all-or-nothing approach to things.

The foregoing should get you started looking in the right direction if your occupational situation presently leaves something to be desired. However, you have to remember that we're taking into account only one portion of your career potential. To do a really thorough job, you must assess your chart as a whole, and this of course is something you can't do strictly by the book because no book can cover all the subtleties of all charts. As each chart is unique, each person has unique potentials in the career area, potentials that can only be described after studying the chart in question. Yet Mars can be an invaluable tool if understood, as it does give you so much information. Once this is digested, you can then proceed to other career indicators—Saturn, Mercury, Venus, the ruler of your tenth house, planets in your tenth house, et cetera—in order to logically and effectively choose the right career for you.

Chapter 4

Energy from Your Environment

Natal Mars is our primary energy source and our chief motivator in life. But it's not the only energy source available to us; nor is it the only thing motivating us in life. There are numerous others, all stemming from our environment. Our parents and relatives, mates and lovers, children, and virtually anyone else whose life is intertwined with ours, however briefly, have the ability to stimulate or interfere with our energy flow and motivate us positively or negatively, according to the interaction between their natal Mars and our natal charts. Furthermore, there are subtle, more or less hidden motivators lurking about in our day-to-day environment. For example, if you own or rent a house or apartment, it has a chart, based on the day it was sold or first occupied; the Mars in this chart affects you. If you work, your place of employment has a chart, based on its date of incorporation or registration; the Mars in this chart also affects you. And the Mars in the chart of your place of residence (a chart that would be based on either the founding or the incorporation date of the city or town) would also affect you.

None of these energy motivators would be as strong as your own natal Mars, although one can have a noticeable impact on you. In some cases it's easy to interpret this impact. For example, there are numerous synastry books on the market that will tell you how interaspects and comparative house placings of Mars affect you. Other times, it's hard to get enough information to make an accurate evaluation unless you're willing to do a lot of digging. After all, how many of you know off the top of your head when your house was first sold or your apartment first rented? And how many of you would know where to find records of your company's incorporation date? This information isn't impossible to obtain, but it isn't normally readily available.

Whether or not it's worth the effort to obtain it depends on you. It's one extra aid in terms of understanding your environment, but if you're coping just fine you may not see any need to go through the hassles involved. And that's ok.

One area where having this extra information can be a big plus, however, is when it comes to choosing a city of residence. The energy of the city can work for you, giving you added energy where your own Mars is weak, or it can work against you, draining you or putting obstacles in your path.

To see how the energies of a city affect you, all you have to do is find the natal Mars position for the city you're interested in and note what house it falls in in your natal chart, and the aspects it makes to your natal planets and angles, if any. (I normally count only those aspects that fall within a two-degree orb.) Marc Penfield is the author of a series of books that give complete charts of cities, states, provinces, and countries around the world, including *Horoscopes of the U.S.A. and Canada* (Tempe, AZ: AFA). If you're at all interested in researching how the energies of your environment affect you, it would be worthwhile to purchase one or more of the Penfield books.

To save time, Table 1 on pages 55-58 shows Mars positions for some major cities. The data in this section comes from a variety of sources, including communication with people who have actually lived in these cities or visited them for reasonably long periods, and from my own experiences with city's energies in my own travels. In other words, these are Mars positions that work for me.

Did you find your city's Mars? If not, choose a city that you've visited or, failing that, one you'd like to visit, and we'll see how the city's energy motivates you.

City Mars in the Houses

First, refer to chapter 1 and look at the sign Mars is in. This will tell you the predominant type of energy generated by the city and into what it's channeled. For example, for Charlottetown, Prince Edward Island, where Mars is in Leo, we'd expect a city where people are fairly active physically and, on a practical level, one where people are encouraged to serve the city or do what the city officials want them to do. The same would be true to one extent or another for all Mars in Leo cities.

Now, turn to the list starting on page 55 and see which of your natal houses the city's Mars falls into.

Table 1. Mars Positions for Major Cities

Mars Sign	Mars City	Mars Degree
Aries	Springfield, IL	01°06'
	Wichita, KS	02°05'
	Savannah, GA	10°57'
	Monterey, CA	11°25'
	Columbus, OH	12°40'
	New Orleans, LA	15°50'
	Aspen, CO	17°55'
	Reno, NY	21°25'
	Harrisburg, PA	27°21'
	Rapid City, SD	28°27'
	Charlotte, NC	29°46'R
Taurus	Pittsfield, MA	03°49'
	Spokane, WA	08°41'
	Oklahoma City, OK	17°50'
	Hastings, NE	21°43'
	Pocatello, ID	22°30'
	Sudbury, ONT, CAN	23°54'
	Natchez, MS	24°59'
	Nashville, TN	29°34'
Gemini	Dawson, YUK, CAN	01°00'
	Hartford, CT	02°31'
	Louisville, KY	08°19'
	Salem, OR	09°24'
	San Antonio, TX	11°03'
	Newport, RI	14°45'
	San Francisco, CA	16°34'
	Saint Augustine, FL	20°11'
	Whitehorse, YUK, CAN	23°06'
	Columbia, SC	24°27'
	Saskatoon, SASK, CAN	26°40'
	Fall River, MA	27°43'

Cancer	Jamestown, VA	01°40'
	Houston, TX	03°07'
	New York, NY	08°24'
	Brigham City, UT	11°34'R
	Toronto, ONT, CAN	16°33'
	Baltimore, MD	17°27'
	Regina, SASK, CAN	24°32'
	Lewiston, ID	25°40'R
	Tulsa, OK	26°29'
	Dodge City, KS	27°54'
Leo	Oakland, CA	00°44'
	Montgomery, AL	05°37'
	Charlottetown, PEI, CAN	07°59'
	Kansas City, MO	09°50'
	Sioux Falls, SD	10°08'
	Brandon, MAN, CAN	11°50'
	Syracuse, NY	12°31'
	Chattanooga, TN	13°16'
	Anchorage, AK	19°29'
	Phoenix, AZ	25°14'
	Toledo, OH	26°13'R
Virgo	Atlantic City, NJ	06°03'R
	Ottawa, ONT, CAN	10°00'
	Jacksonville, FL	13°34'
	Philadelphia, PA	14°20'
	Hamilton, ONT, CAN	16°30'
	Juneau, AK	17°24'
	Raleigh, NC	19°04'R
	Daytona Beach, FL	20°00'R
	Frankfort, KY	24°54'
	Montpelier, VT	26°30'
	Chicago, IL	27°58'
	Austin, TX	28°45'

Libra	Edmonton, ALTA, CAN	01°51'
	Huntington, WV	05°57'R
	Fargo, ND	06°53'
	Birmingham, AL	07°07'
	Akron, OH	08°19'
	Buffalo, NY	10°09'
	Tucson, AZ	24°20'
	Sacramento, CA	26°47'
	Dallas, TX	29°52'
Scorpio	Las Vegas, NV	15°02'R
	Dover, DE	17°16'
	Mobile, AL	18°47'
	Little Rock, AR	23°02'
	Twin Falls, ID	24°14'R
	Rochester, MN	26°33'
	Fairmont, WV	28°29'
Sagittarius	Charleston, SC	01°11'
	Trenton, NJ	02°10'
	Victoria, BC CAN	09°36'
	Santa Fe, NM	12°38'
	Long Beach, CA	15°06'
	Cleveland, OH	17°15'
	Wilmington, DE	18°45'
	Des Moines, IA	19°46'R
	New Haven, CT	21°42'
	Pittsburgh, PA	23°55'
	Calgary, ALTA, CAN	25°16'
	Alexandria, VA	28°27'R
Capricorn	Halifax, NS, CAN	03°04'R
	Fort Wayne, IN	05°19'
	Cincinnati, OH	07°34'
	Portsmouth, NH	08°37'
	Topeka, KS	11°40'
	Augusta, ME	12°17'

	Rock Springs, WY	15°59'
	Los Angeles, CA	17°49'
	Grand Rapids, MI	27°44'
	Price, UT	24°42'
	Honolulu, HI	28°11'
	St. John's, NFLD, CAN	29°20'
Aquarius	Wheeling, WV	01°02'
	London, ONT, CAN	02°20'
	New Bedford, MA	05°26'
	Portland, OR	06°31'
	Denver, CO	07°04'
	Quebec City, QUE, CAN	16°09'
	St. John, NB, CAN	23°30'
	Asheville, NC	25°56'
	Missoula, MT	26°24'
Pisces		
	Fredericton, NB, CAN	04°52'
	Miami, FL	08°05'
	Montreal, QUE, CAN	10°14'
	Mankato, MN	15°43'
	North Platte, NE	16°18'
	Oshkosh, WI	22°29'

City Mars in Your First House

This may encourage you to be more ambitious than you'd otherwise be. It can also encourage you to rush, so there may be accidents while you're living or visiting here. New beginnings would almost certainly be a theme in your life.

City Mars in Your Second House

This can help you increase your income, but it might also encourage you to spend your money as fast or faster than you make it. It would certainly make you more ambitious for money than you might otherwise be. It might encourage you to become involved in a business related to machinery or another Mars-ruled kind of work.

City Mars in Your Third House

Restlessness would be increased and you might take many short trips within the city or to neighboring cities. Debates or even arguments could be more frequent than they otherwise would be. You could become motivated to take action to improve you neighborhood.

City Mars in Your Fourth House
This could increase the possibility of fires or injuries in your home. Somehow this city reminds you of things you don't want to be reminded of, causes you to dwell on painful memories, or otherwise upsets you. This is one instance where you'd probably be wise to look at relocation.

City Mars in Your Fifth House
This can encourage you to make impulsive romantic decisions. It intensifies your sex drive, particularly if it's in aspect to your natal Mars. If there are stress aspects between this Mars and your natal planets, this can be a dangerous city for you. Problems seem to be most common for people with cardinal-predominant charts.

City Mars in Your Sixth House
This doesn't do wonders for your temper! On the other hand, it makes you more energetic than you might otherwise be in the work area, and helps you accomplish things you might otherwise never get done.

City Mars in Your Seventh House
A lot depends on what aspects, if any, Mars makes to your natal planets. If these are positive, it can help make your social life more enjoyable and can bring lots of parties, dates, and nights on the town. If they're not, then this position increases the possibility for unpleasant confrontations, lawsuits, and marital troubles. If aspects are stressful, you'd be wise to look at relocation.

City Mars in Your Eighth House
There may be more trouble than otherwise with joint finances. You may be more inclined to attend occult activities. Or you could become motivated to be active in some sort of corporate business activity.

City Mars in Your Ninth House
Be careful where legal matters are concerned, as these will be more difficult for you than otherwise. There's an increased danger of religious fanaticism, too. Or you could get Involved in some sort of political activism.

City Mars in Your Tenth House
It will stimulate you to become more ambitious. The career area is most affected by this, and changing jobs or starting new career activities becomes a theme.

City Mars in Your Eleventh House
You'll get plenty of social invitations, increase your casual contacts, and be motivated to take energetic action with friends. Your feelings may become more revolutionary or militant than they'd otherwise be.

City Mars in Your Eleventh House
This tends to make you more impulsive than otherwise, and increases the chance of you being confined to a hospital for surgery or due to an accident. It can also make you more secretive.

Aspects to City Mars

In terms of aspects, use the following guidelines:

- Conjunctions energize and emphasize what they touch.
- Sextiles and trines can both be considered helpfully energizing; however, the sextiles suggest opportunities that are more dependent on your motivation (as shown by your natal Mars) than are those of the trine.
- Squares and oppositions tend to block or misdirect your energy flow. With squares, your motivation is not in keeping with what the city deems to be in its best interest. With the opposition, individuals, rather than the tempo of the city, are the source of difficulty. Oppositions tend to draw you to people who either drain you or actively oppose you.
- Semi-sextiles show minor irritants that grate on involved planets. Quincunxes also grate, but where the semisextile normally triggers temper, the quincunx triggers brooding. It can, as a result, increase your susceptibility to certain illnesses. Remember, use no more than a two-degree orb. All interpretations that follow refer to the city's Mars in aspect to your natal planets.

City Mars Conjunct Your Sun
This increases your impulsiveness and may therefore contribute to faulty judgment. You often feel the city is somehow holding you back. Male bureaucrats can be a particular problem. On the plus side, Mars increases your energy level.

City Mars Sextile/Trine Your Sun
This makes people in high places a bit more receptive to you than otherwise. It also improves your health, increases your energy level, and motivates you to become involved in some sort of energetic physical activity.

City Mars Square/Opposition Your Sun
This strengthens any impulsive or aggressive tendencies in your natal chart. Sometimes you feel that others are opposing your plans without good reason.

City Mars Conjunct Your Moon

This makes you more emotional than you might otherwise be, and in extreme cases can contribute to instability. It increases the likelihood of limitations in the career and domestic spheres.

City Mars Sextile/Trine Your Moon

You may be more involved with women that you might otherwise be (not necessarily romantically). This increases enthusiasm as a general rule, although the sextile is more changeable in mood and more passive than trine.

City Mars Square/Opposition Your Moon

This increases the tendency for impulsiveness, tactless outbursts, speeding, and being accident prone when found in a natal chart.

City Mars Conjunct Your Mercury

This greatly increases argumentative and impulsive tendencies, so if these are already strong in your natal chart, chances are this city will be hazardous to your health, emotions, and success potential. Mars attracts you to intellectual debates and disputes in this position as well.

City Mars Sextile/Trine Your Mercury

You can think faster than you might otherwise. This also encourages you to answer correspondence and return phone calls more quickly, and it helps you assert yourself and communicate your ideas in a positive manner.

City Mars Square/Opposition Your Mercury

This increases the likelihood of annoyances connected with correspondence or communication, and can contribute to tension to your life. It makes it harder than it might otherwise be to control your temper.

City Mars Conjunct Your Venus

This can make you more attractive to the opposite sex. It generally brings more harmony than discord into your life unless one or both planets are under severe stress. It's also an aid in terms of public relations in general.

City Mars Sextile/Trine Your Venus

You're motivated to find fun and enjoy yourself, so in a chart where ambition isn't strong, this can encourage you to ignore or neglect this work. If your natal chart shows you to be a highly emotional, romantic person, these aspects can get you into trouble in the romantic area by encouraging a tendency to go overboard. Normally, these aspects will make you more tactful than you otherwise would be.

City Mars Square/Opposition Your Venus

This encourages carelessness in the financial area and hypersensitivity in the emotional area. The slightest thing tends to hurt your feelings.

City Mars Conjunct Your Mars

This sharpens (and sometimes loosens) the tongue. It also increases your initiative and can motivate you to become involved with tools or other Mars-ruled things.

City Mars Sextile/Trine Your Mars

In extreme cases, this can contribute to faulty judgment, encouraging you to rush or make snap decisions. Normally, however, it motivates you to take intelligent action. It also raises your vitality and energy levels.

City Mars Square/Opposition Your Mars

Things don't move fast enough for you here. A tendency for anger, impatience, and impulsiveness is stimulated.

City Mars Conjunct Your Jupiter

This encourages over-optimism in financial matters and extravagance in spending. It increases your enthusiasm, and also the danger of religious fanaticism. If your natal chart points to financial difficulties, or if this conjunction falls in your second, fifth, or eighth house, this might not be the best place for you to live.

City Mars Sextile/Trine Your Jupiter

This helps things run smoothly in the work area, improves your timing, and helps you settle legal matters in a satisfactory manner.

City Mars Square/Opposition Your Jupiter

This tends to attract unexpected expenses and can drain your finances. You tend to take too much for granted. In extreme cases it can encourage unethical behavior in work or financial areas.

City Mars Conjunct Your Saturn

This could be a frustrating city to live in. Although this aspect motivates you to take on hard physical work and increases your drive and ambition, it is often at the expense of your emotions. You can become harsh, cold, and difficult to get along with. Relationships in general tend to be difficult. This is one instance where you might be more content elsewhere.

City Mars Sextile/Trine Your Saturn

This helps you get results from your efforts and make progress in the work area. It's also a good influence on health and energy levels.

City Mars Square/Opposition Your Saturn

You'll have to work harder than might otherwise be the case. This might cause you to experience minor conflicts or to try to go in two directions at once. You may well feel you want to get away from this city.

City Mars Conjunct Your Uranus

This aspect is exciting at first, but eventually it can get to you by causing trouble in the involved house to the extent where you may decide to toss away much of what you've worked for and try your luck elsewhere. On the plus side, it can encourage you to take more initiative in joint finances. On the minus side, it can have an adverse effect on your health and/or nerves and contribute to sudden-onset difficulties. This might be a fun place to visit, but I wouldn't advise you to live there!

City Mars Sextile/Trine Your Uranus

This increases your determination and resourcefulness and could help make your intuition more reliable.

City Mars Square/Opposition Your Uranus

This tends to make you more egotistical than you otherwise would be. It also increases carelessness and could thus contribute to unexpected accidents, as well as increasing the likelihood of your life being disrupted by circumstances beyond your control. This is not recommended as a place to live if you want good control of what goes on in your life.

City Mars Conjunct Your Neptune

This can be a confusing place to live. The conjunction stimulates secretive tendencies and motivates you to have dealings with religious and occult institutions, hospitals, et cetera. If your natal chart is over-abundant in fire or air (or cardinal placements), this might be a good bet, as it will add balance to your chart.

City Mars Sextile/Trine Your Neptune

Occasionally this increases a tendency to look at the world through rose-colored glasses and get carried away emotionally. Normally, however, it encourages a positive emotional outlook. This can increase the likelihood of getting positive guidance from your intuition in terms of work and other matters.

City Mars Square/Opposition Your Neptune

Be extra careful if your natal chart contains additional stress aspects to Neptune; water damage or emotional problems could occur. This increases the tendency to suppress your anger and strengthens your subconscious as a motivating force in your life.

City Mars Conjunct Your Pluto

This increases your aggressive tendencies and intensifies your willpower. It's not a good place for you if there's civil unrest there, though, as you could be caught up in this whether you want to be or not.

City Mars Sextile/Trine Your Pluto

These aspects greatly increase your assertiveness and your ability to put forth sustained effort. Usually these aspects help you get what you want; at the very least they increase success potential. However, if you're already inclined to be stubborn, this tendency will also increase, and compromise may become more difficult for you.

City Mars Square/Opposition Your Pluto

This increases aggressive tendencies and tends to make you more provocative than otherwise; hence it increases the likelihood of arguments. It also stimulates your temper and increases the likelihood of impulsive action.

City Mars Conjunct Your Ascendant

This increases impulsiveness, and in extreme cases it can contribute to a serious problem with aggression. Normally, though, it merely increases any workaholic tendencies you might have.

City Mars Sextile/Trine Your Ascendant

This makes it slightly harder for you to relax. You're motivated to start many projects and tend to be especially attracted to things that require physical activity. These aspects greatly increase your physical energy level.

City Mars Square/Opposition Your Ascendant

This motivates you to be impulsive in your actions. You tend to be in too much of a hurry for your own good, which can increase the likelihood of accidents. In extreme cases this could attract some sort of physical violence.

City Mars Conjunct Your Midheaven

This increases the likelihood of quarrels in general and conflict with your parents or bosses in particular. It also increases the likelihood of competition in the career area. Unless you thrive on challenge and are strongly independent, this is probably not the best place for you to live.

City Mars Sextile/Trine Your Midheaven

This could attract help from your parents, a boss, or some older male; however, you may be expected to do something for your helper in return. These aspects tend to make you more dynamic and more ambitious in terms of career and status.

City Mars Square/Opposition Your Midheaven

This is a tiring, draining influence. It increases the likelihood of conflicts in the career area and in your private life. For some inexplicable reason, it also seems to attract car trouble and unexpected transportation difficulties.

Note that the city's Mars can't act on your personality to make you into something you're not. If, for example, you're afraid of romantic involvement, the city's Mars might make you less fearful, but it won't suddenly propel you into romantic bliss; you'll still go fairly slowly and carefully in romance, but you won't be so closed to opportunities. Conversely, a city's Mars doesn't "give" you problems unless you're predisposed to those problems in the first place. So your natal chart shows you to have a wicked temper, don't say it's the city's fault. Your locality may well aggravate your temper and may even be the catalyst, but it's not the whole problem. The bulk of the problem is your low annoyance threshold. Rather than blame your environment, therefore, you should do something constructive about this. Meditate, jog, seek hypnosis, whatever. Just try not to spend your time griping about what's wrong without seeing what's right. And obviously, if you feel nothing is right, it's time to move on. Sometimes this is truly the best solution. But remember, when you move, you'll take that wicked temper with you. You can minimize it by going to a more congenial place, but the only thing that will get rid of it is constructive effort on your part.

The chief thing to keep in mind when planning a move is that you can minimize weaknesses or emphasize strengths if you choose your location wisely. But you can't escape from yourself.

66/The Mars Book

Chapter 5

Mars and Your Luck

Mars and your luck??? I hear disconcerted mental "voices" in the background as you read that title. "What's Mars got to do with luck?" you say. "Everybody knows that Jupiter rules luck!"

Yes, everybody knows that—or at least everybody is taught that, but I somehow don't believe it. Why? Well, first of all, I don't believe in luck in the sense of goodies just falling from the blue into your lap. I've never seen it happen that way. To be lucky, you've got to create opportunities for good things to happen. The good thing you want can be a lottery win; a trip; a tall, dark, handsome, rich man who's madly in love with you; or whatever. To get it, you still have to buy the lottery ticket, enter the contest, go out and meet people, et cetera. And that means taking action of some sort. What? You don't believe me? You say maybe dear old Uncle Ed will give you the winning lottery ticket for your birthday or Prince Charming will inexplicably drop in at your workplace and fall madly in love with you at first glance? Maybe, but you still aren't getting a totally free ride. If you meet Prince Charming at work, that means that somewhere along the line you've taken action to find yourself a job and then, on the appointed day, to be there! And if dear old Uncle Ed remembers your birthday, it's no doubt because you've taken action to keep in touch with him and let him know he's your favorite uncle. Maybe you even told him it was your birthday. But somehow you took action. And action, as everybody knows, is under the rulership of Mars.

Okay, so luck doesn't come without some sort of action on your part. What, then, is luck anyway? My definition of luck is being able to take action when opportunity comes your way, and knowing how to distinguish good opportunities from bad. In other words, I see luck as good timing.

To see what sort of timing you have and what sort of opportunities are most likely to present themselves to you, we'll look at Jupiter by sign and house placement, look at Mars by sign and house placement, and finally at the aspects between Mars and Jupiter in terms of the natal chart. When we've done that, you should be all set for a winning streak. Ready?

Jupiter by Sign and House

First, let's look at Jupiter by sign and house. This shows us your attitude toward growth and expansion and how opportunities tend to present themselves.

Jupiter in Aries

You may not give as much serious thought to your goals as you should; as long as your time is being spent enthusiastically rushing from one hot trend to the next, your luck tends to be hit-and-miss, with the misses invariably costing you what you gained on the hits! Once you have goals for yourself firmly in mind, your timing invariably improves.

Opportunities may appear in the guise of a new chapter in your life; that is, they may frequently appear shortly after a move, job change, change in marital status, et cetera. Helping others often leads to lucky breaks. Your patience may be tested to the fullest as you wait for these breaks to materialize, however. Sometimes other people hold you up (and drive you up the wall) by not taking the initiative when you feel they should and, as a result, being in your way when you're ready to open a new door. You may, when this happens, decide to take the initiative for them—or even to knock them out of the way to get ahead. Invariably this is when your timing goes awry and "bad luck" comes into the picture.

Jupiter in the First House

This is similar to Jupiter in Aries. Unless Jupiter is in a fixed sign or in aspect to Saturn or Pluto, self-discipline is not your strong point and timing may be off as a result. You enjoy being free so you can have as many kinds of experiences as possible. You want your piece of the pie of life to be a large one—preferably a la mode! You therefore tend to do things to excess at times. You don't like anything to be small; your attitude toward gains is "all or nothing," and "a bit" is nothing. Unless Jupiter is in Aries or Sagittarius, or conjunct Mars, you aren't be as inclined to knock people out of your way to get what you want.

Jupiter in Taurus

The first thing you have to learn is generosity; this will pay off for you in important ways. But you must be sure you're practicing true generosity as paying money to assuage your guilt or trying to "buy" people with gifts, favors, or anything else will backfire. Also, Jupiter in Taurus can cause you to behave in a possessive manner. This of course is the negative side of this placement and as such is not the side you should concentrate on using if you're trying to be successful and lucky! The good side, the side to develop, stresses investigating potential in general, and paying special attention to artistic, agricultural, or horticultural projects. Opportunities for lucky happenings can also take the form of trips, reunions, marriage, or virtually any circumstance in which you spend money on others without thought of gains for yourself.

Jupiter in the Second House

This placement has always been considered the mark of someone who easily attracts money. This is true, but it can also cause money to be easy-come, easy-go if you're not careful. In fact, I've had more than one client with Jupiter here who's gone bankrupt! Sometimes there's a funny tendency to buy the best on the market in terms of whatever you're interested in and then sit around and admire it rather than use it. Hoarding is occasionally seen, too. These tendencies would need watching.

Jupiter in Gemini

You have the best luck with projects and activities of fairly wide scope. Art, journalism, and broadcasting are areas in which lucky breaks can occur. Romance is another area. Siblings and cousins can sometimes also open doors for you. A tendency to work too hastily is the prime hazard with this placement.

Jupiter in the Third House

This is similar to Jupiter in Gemini, with neighbors, relatives, and siblings often being a source of opportunities for you. You're very curious and probably go through periods of intense questioning of people; when you do this, it's because you're searching for some sort of answer—perhaps even an answer to a question you're not even aware you formulated. But you don't want advice rammed down your throat. You want to be free to make up your own mind about what to do with your life. Normally this is a plus, but occasionally it can be a minus, for if someone gives you a hard-sell talk about something—even if the something is truly a golden opportunity—you may become contrary and reject it out of hand. This is particularly true if Jupiter is in a fixed sign or under stress from Mercury or Uranus. Finally, this position quite often brings good opportunities in connection with travel or moving.

Jupiter in Cancer

Your most obvious source of lucky opportunities is in connection with your early home life, or perhaps your mother. A less obvious source would be loved ones in general. But there are also other areas. Hobby-businesses involving creative work done from your home, metaphysical pursuits, and sales are some examples. Sometimes this placement gives a sort of misguided craving for respect can make you assume—wrongly—that if you please others, they'll reward you. Unfortunately, it doesn't always work this way, so you're better off to develop your self-respect and worry less about what others are going to think if you disagree with them.

Jupiter in the Fourth House

This is similar to Jupiter in Cancer. Opportunities tend to be connected with your upbringing, a parent, or perhaps your childhood home. You may eventually own a great deal of land or may sell land you own at a profit. You might even be involved in a real estate career. Anyway, buying and/or selling land is liable to be a theme.

Jupiter in Leo

In terms of providing opportunities, this placement has few equals. But it's not entirely without drawbacks. For example, it may tempt you to brag when you should be modest, to rebel against common-sense precautions, or to seek attention from the wrong people. These tendencies can put more frustrations than opportunities into your life in the long run. Showing healthy anger over real unfairness or incompetence is okay, but blowing your top in hopes of being given your own way just to shut you up isn't a good idea. Good opportunities can come your way in connection with creative projects (especially film or theater), dealing with young people, or counseling.

Jupiter in the Fifth House

This is excellent for creative pursuits as a rule, unless it's conjunct Saturn. With this placement, you need to be free to find your own way in life. Like your Jupiter in Leo counterpart, you don't like anyone trying to restrict you, even if they're trying to protect you from potential danger. You may enjoy sports very much; if so, these could be a source of lucky breaks. You'd probably have the best luck with outdoor sports, as psychologically you tend to feel more free in these than with indoor sports.

Jupiter in Virgo

The greatest number of opportunities in your life are connected with finding practical uses for your idealism. Writing, research, and activities connected with appliances or equipment that can save you time or money are good paths to growth and luck. On the minus side, there's sometimes a tendency to reject new people or things before you've really given them a fair chance. The temptation to make mountains out of molehills also has to be avoided if you want opportunities to knock.

Jupiter in the Sixth House

Usually less inclined to make mountains out of molehills, this tends to make you growth-oriented and anxious to make the most of your potential. For this reason, you need to find work that's meaningful to you. If you're bored with your work, you become frustrated, and this can make it difficult for you to successfully evaluate your opportunities for one reason or another. With Jupiter in a fire or air sign, for example, this frustration tends to make you impulsive and you leap at anything that looks even remotely like an opportunity, with the result that you may be too busy chasing rainbows to see really good opportunities when they appear. With Jupiter in an earth or water sign, on the other hand, boring work can eventually anaesthetize you so that you're not alert when opportunities come.

Jupiter in Libra

Your opportunities tend to be connected with relationships, psychological growth, and/or aesthetics. They tend to surprise you when they come because you sometimes don't take things seriously enough and are later amazed when you find out that what you thought was a trivial comment or action has elicited a strong response. With this position, there's sometimes an inherent conflict between your desire to nurture others and to help them grow with you, and your desire to protect yourself from unpleasantness and discord. Being dependent of others (or encouraging their dependency on you) or avoiding necessary confrontations—and, where necessary, severance of ties—can keep lucky breaks from materializing.

Jupiter in the Seventh House

This is similar to Jupiter in Libra. In this case, you try to grow by involving yourself in close relationships. You therefore tend to find that relationships have a marked effect on your luck for better or worse (depending on the people you choose to get close to). Sometimes this placement is indicative of a mate who has conflicting desires for nurturing the mate and putting the mate first and protecting his or her interests by looking out for number one.

Jupiter in Scorpio

Your best opportunities tend to be connected with expanding your resources, making profitable purchases, finding bargains, and studying the occult. This placement is normally a sign that a great many opportunities await you in life. To make the most of these, however, you have to accept the fact that you will occasionally need help from others. So you have to learn how to handle people in a sincere, non-overwhelming manner. You also have to watch out for the resentful streak that sometimes comes with this placement.

Jupiter in the Eighth House

This is similar to Jupiter in Scorpio. It may give luck in the areas of medicine, psychology, and/or the occult. If you share what you have with others, they'll share with you and help you progress. In general, others do things that bring opportunities your way, but it's what you do yourself that determines your success or lack of it. Resentment is very occasionally a problem with this placement, but not as common as with Jupiter in Scorpio.

Jupiter in Sagittarius

This placement gives you a great deal of inner vision, although it doesn't always guarantee you'll be able to get others to see what you see. "Going with the flow" is generally the best way for you to find unexpected luck opportunities. Paperwork, teaching, and writing letters all tend to bring opportunities into your life.

Jupiter in the Ninth House
Sometimes inclined to be a trifle too self-righteous, which could nip some opportunities in the bud, you tend to be fascinated by anything new that is in harmony with your philosophy of life. Exploring what appeals most strongly could bring you luck.

Jupiter in Capricorn
Your biggest source of opportunities is your attitude toward your career and your ability to realize when it's time to expand or move on. You can benefit yourself most by helping society make necessary changes and by expanding your own social horizons. Teaching and studying—particularly in an area that involves math or numbers—can bring you luck. Fighting change or doing little "tit for tat" scenes will slam doors in your face.

Jupiter in the Tenth House
Similar to Jupiter in Capricorn, opportunities very often revolve around caring for or being in charge of others in some way. If you equate luck with professional success, this is a good placement to have; usually it makes for considerable progress in the career area and a position involving some prestige. Often this position doesn't come until after age thirty; in fact, it most commonly comes between ages thirty-five and forty-five. There is a problem with this placement in that it can encourage smugness and a swelled head. If bad luck assaults you, these may be the reasons.

Jupiter in Aquarius
No doubt there will be several prime opportunities in your life. Often you accidentally step into scenes that are rife with opportunity. Luck for you is connected with getting involved in large projects, particularly those connected with science or with the betterment of humanity on some scale. Luck can also be connected with marriage or other committed relationships. Your sense of timing can be either a boon or pitfall. You do tend to be "out of sync" fairly often—sometimes way ahead of the times, sometimes busily balking at the inevitable instead of preparing for it. This of course delays success and doesn't make for smooth sailing. But it doesn't doom you either!

Jupiter in the Eleventh House
Similar to Jupiter in Aquarius, you should have good luck with friends throughout most of your life and probably would be both an idealist and an optimist.

Jupiter in Pisces
You have tremendous opportunities to discover things that are overlooked by others. Luck can be found in the areas of acting, dancing, writing, or any art form. However, you have to practice and/or train if you're to realize your fullest potentials. Sometimes people with this placement enjoy dreaming more than doing; as a result, they wind up being mediocre when they could have

been great. Another stumbling block is indecisiveness. There's a tendency to let circumstances decide things for you, often in a way that leaves you "out of luck."

Jupiter in the Twelfth House
This influence tends to be inward-looking and more concerned with self-development than acclaim. You're a very kind person whose kindness may be rewarded and may provide opportunities for success even though you're not looking for success and are perhaps actually opposed to it if it would disrupt your private life.

Jupiter affects timing to some extent regardless of its relationship to Mars. I've noticed that Jupiter in a fire sign tends to be the quickest to answer the door when opportunity knocks, followed by Jupiter in earth signs, Jupiter in water signs, and Jupiter in air signs. I don't know why this is. It could have to do with fire and earth being more outer-world oriented and Jupiter in air and water being more inner-world oriented. Or it could be that fire and earth are less inclined to be swayed by other people's opinions than are water and air. Of course, any aspects between Jupiter and Mars could change the picture, speeding things up or slowing things down accordingly.

Mars and Opportunity

Mars shows you how much energy you have at your disposal for taking advantage of lucky breaks, what you do to put yourself in a position where opportunity can find you, and in which areas your motivation is highest.

Mars in Aries
You're highly motivated in the intellectual area, particularly in the field of psychology. Electronics and home computers also intrigue you, although in these fields a lot depends on your available funds. Being curious, you tend to be an avid reader, and at times a book read simply because nothing else is handy later come to your aid by providing needed knowledge in order to take advantage of a lucky break. Conversations and meetings—both planned and spontaneous—tend to alert you to opportunities that excite you. Speculative topics such as religion, politics, and astrology are particularly apt to motivate you. You may get lucky breaks as a result of writing to newspapers or magazines; in fact, the field of publishing in general can be a source of good opportunities.

Mars in Taurus
Your motivation is highest when it comes to practical opportunities. Your style is to work hard and play hard. Art, music, and working with growing things can steer you toward success provided you don't close your eyes to new trends that might affect your work. (You are, at times, a trifle negative about change.) Romance can provide tangibly good opportunities at times. Building construction or decorating can also be a lucky theme in your life.

Mars in Gemini

Your motivation tends to be strongest when you're pursuing things or ideas other people find odd. Travel can bring lucky breaks. Your occupation may also be a good source of opportunity, particularly if it involves writing, public speaking, or mechanical pursuits. Marriage could bring a major change in your motivation or choice of pursuits, for better or worse.

Mars in Cancer

Working hard around your home is one way to find opportunities for personal growth. It doesn't really matter whether your work is actually connected with your home or merely done in your home; if your motivation is high, the opportunities will come. Cooking, delivery work, and the business world in general can also provide you with lucky breaks. In general, the less glamorous the field, the more opportunities you'll find, as you're most highly motivated when you see unfilled needs or people needing protection. This may often mean succeeding because you take on work nobody else wants to do.

Mars in Leo

Your motivation is high when you're contemplating large scale action—trips to foreign countries, making a hit on Broadway, harnessing solar energy, etc. Lucky breaks tend to come when you slip off to do your own thing rather than when you follow the crowd. However, crowd scenes can sometimes provide small opportunities, provided there aren't too many bureaucratic maneuvers to contend with. Herb gardening, flower arranging, acting, dance, and astrology are all areas where your motivation would tend to be high even though the scope of your work may not initially be far-reaching.

Mars in Virgo

Working hard comes naturally to you. Your job, relocation and/or work transfers can lead to some of your best lucky breaks. Less important opportunities can come as a result of seemingly trivial things like going for a drive, getting a new refrigerator, shopping for groceries, having your computer repaired, et cetera. The key here is that your opportunities tend to come in connection with serviceable things and practical activities rather than from involvement with luxuries or highly "iffy" projects. Coworkers, editors, and professional people in general tend to motivate you by giving you encouragement.

Mars in Libra

You're motivated to be a success in the work world and thus are willing to relocate, work long hours, or even work for a disagreeable boss as long as you see a promotion, raise, or some sort of prestige out there waiting for you. The areas in which your lucky breaks fall tend to vary according to your definition of luck and what you want out of life. Romances—even very short-lived ones—are common door-openers for you. Participating in social events and working in travel agencies, salons, or getting involved in dance or exercise programs can bring opportunities.

Mars in Scorpio

You're highly motivated to see everything there is to see and do everything there is to do; for this reason, travel, reading and expanding your circle of business and social contacts would tend to strongly appeal to you. Romance, pregnancy, and caring for others are activities that frequently open doors. Psychology can be a good source of opportunity for you, too. I find that sometimes this position has a way of bringing lucky breaks at times when you're either too tired or too busy to fully appreciate them.

Mars in Sagittarius

Unless an activity has tremendous potential to improve your life, your interest in it passes quickly and your motivation remains low. Trips to little-explored foreign places and opportunities to teach or do research in new, wide-open fields would be typical of the sort of activity that strongly motivates you. On a smaller scale, social activity involving shared creative or intellectual interests, or meeting people who somehow expand your mind, can lead to lucky breaks. So can romance. Fads, especially those of a recreational nature, can provide you with good opportunities, as can any action on your part that's purposely designed to remove restrictions from your life.

Mars in Capricorn

You're highly motivated to be successful in the career area. Social and recreational interests, therefore, tend to take a back seat to work. You might unconsciously say no to lucky breaks if they come in these two areas because you have work planned and are reluctant to reschedule it. You must remember that it is possible to successfully mix work and pleasure. Opportunities can be linked to friendships as well as to your work.

Mars in Aquarius

To be quite frank, it's impossible for me to generalize about what motivates you. Mars in Aquarius, taken on its own, just doesn't have a typical, consistent motivator. What really motivates one person is apt to be the same thing that causes someone else to turn up their nose in disgust. Furthermore, what motivates you now may irritate you or just plain leave you cold a month from now. Trips, social activities, and club involvements can all bring lucky breaks, provided you're keen on them, but there's no way to be more specific as far as I can see.

Mars in Pisces

Psychology motivates you, as does virtually anything creative or occult or anything that can be done now and forgotten about for a long time. By the latter, I mean one-shot, intensive projects—things that may involve hard work but give long-term (preferably permanent) results once completed. Motivation flags in areas where the work is repetitive and has to be done regularly to stay caught up. You meet luck in the form of people who either help you do things you can't do yourself due to lack of skill or who help you pay for things you truly need. Lucky breaks can also come in the career area, but these are not always taken advantage of due to a fear that career success will detract from harmony on the home front. Productive hobbies are a good source of opportunity, particularly those connected with beautifying your home, exploring your subconscious in order to remove failure programs, or doing anything that helps you become more self-sufficient. Theater and film work, social service work, and hospital work are also good areas to explore.

Mars in the First House
This is a sign of a high level of motivation, so undoubtedly there'll be a great deal of activity in your life. You're particularly good at fighting for your rights, and in that sense you make your own luck. Your physical energy level is liable to be quite high, enabling you to take action when opportunities come your way. However, if Mars is in Cancer, Libra, or Pisces, your energy may be more mental than physical and you may have to push yourself a bit in order to avoid the tendency to let circumstances decide for you.

Mars in the Second House
You're motivated to be somebody special in some way and to own nice things. This can be good or bad, depending on whether you equate luck solely with money and prestige or whether you're willing to settle for less tangible sorts of luck. You're willing to work to establish yourself so that opportunity doesn't overlook you, which is good. However, if money is your only goal, you may pass up some potentially worthwhile activities and thus short-change yourself.

Mars in the Third House
Your motivational level is good, and the tempo of activity in your life is apt to be fairly brisk. You sometimes make your own bad luck by trying to coerce others into believing as you do or doing what you say. This generally happens when you're feeling insecure for some reason; it's a form of over-compensation. It can result in you having the doors slammed in your face even though your skills or ideas are worthy of consideration, so watch this tendency. Your best opportunities are often connected with writing and other sorts of mental work—the harder, the better. For this reason, a good education would be a great asset.

Mars in the Fourth House
You're often motivated by hidden or subconscious factors that are triggered by environmental stimuli and stir your instinctive defense mechanisms. Lucky breaks are apt be connected with your home and/or birthplace, and a great deal of your activity may be connected with these (sometimes it's getting away from them that leads to lucky breaks). To a lesser extent, lucky breaks may be connected with serious involvement with community recreation or promoting local businesses. Career activities tend not to offer too many opportunities as a rule, either because you're a rather private person or sometimes because of a militant or egotistical approach that turns people off.

Mars in the Fifth House

A need to be yourself is what motivates your actions more than anything else. If you have children or work with children, they can be a source of lucky breaks. Interest in theater, fitness, sports, or alternative lifestyles might also be good sources. If opportunities aren't coming your way, it may be because you lack discipline; if you don't want to do something, your tendency is to procrastinate in hopes the need to do it will just go away. Unfortunately it won't, and if you procrastinate for too long, you'll only wind up being unprepared when opportunity knocks.

Mars in the Sixth House

You're reasonably energetic, a hard worker, and are motivated by praise. There can be a problem with this placement; since Mars here says the primary use of your energy is to serve others, these others can keep you from getting the breaks you deserve if you don't choose your "masters" wisely. Coworkers, relatives, your mate—all can enslave you or conspire against you (consciously or otherwise) if they feel your progress wouldn't be to their advantage. You have to be aware of this and should refuse to let anyone become too dependent on you in any capacity. Opportunities for you tend to come in the areas of health, writing, and/or service or support staff positions.

Mars in the Seventh House

Tension motivates you. This can be good or bad, depending on whether you use your energy to resolve problems or to get back at those whom you feel are causing your bad luck. Harmonious relationships can be good sources of opportunity for you. Discordant relationships can also lead to positive opportunities if they help you determine your priorities and understand what motivates you. Be careful of childish "tit for tat" behavior, though, which is counterproductive.

Mars in the Eighth House

Your motivation can be very material or very metaphysical. Your lucky breaks can be either very big or very small, depending on how ambitious you are and what you choose to do with your life. Conflicts very often actually turn out to be opportunities in that they force you to reexamine your wants and bring them more in line with your needs and abilities.

Mars in the Ninth House

Your motivation is strongest when you're doing creative intellectual work—teaching, programming, devising games, et cetera. You put almost all your energy into expanding your mind, so lucky breaks tend to take the form of finding information. Opportunities tend to be connected with law, metaphysics, and religion.

Mars in the Tenth House

Your motivation to gain ego gratification is high. You are extremely ambitious, and want to achieve success, so you work hard and create your own lucky breaks for the most part. Conflicts with bosses may arise, and there can occasionally be conflicts with parents. These people may resent your success or feel threatened by your efforts to get ahead. They may also try to prevent you from having further opportunities to succeed, particularly if they feel you're making them look dumb or lazy, or that you're after their job. Watch out for that and your luck should be good.

Mars in the Eleventh House
You're motivated to formulate goals for the future. These basically involve ego-enhancement of some sort and you use your energy to pursue them. Physical activity involves friends or clubs, and membership in scientific, psychological, parapsychology groups or any activity that allows you to make contact with important people can be a source of lucky breaks.

Mars in the Twelfth House
Your motivational level is erratic. You tend to give up if you think people aren't handling your ideas properly or aren't appreciative of your efforts. Because you don't like surprises, you often feel irritable rather than pleased when lucky breaks present themselves. You want to succeed all by yourself; you tend to alternately resent offers of help (seeing them as interference), and complain about how much you have to do. In terms of opportunities, they tend to be best—and best appreciated—if you work alone. If possible, choose a creative field.

Mars-Jupiter Aspects

Mars-Jupiter aspects show your sense of timing and your ability to be in the right place at the right time.

Mars Conjunct Jupiter
This shows an excellent sense of timing in connection with virtually any kind of physical activity. It gives strength, stamina, good motivations, the ability to project self-confidence, and normally, a belief in yourself. It usually makes for a very exuberant sort of personality, so occasionally you may "jump the gun" or not realize how hard something is before you tackle it.

Mars Semi-sextile Jupiter
This is an "almost-but-not-quite" influence in terms of timing. Depending on sign and house position, you may be either a little too fast or a little too slow. You have a sort of teaching skill in that you enjoy introducing people to what interests you and will spend much time explaining it. What you aren't interested in, you procrastinate about. You can also be impractical at times or can try to start at "C" without even looking at "A" and "B".

Mars Sextile Jupiter
This is normally very positive in terms of timing, and your physical energy tends to be high, your motivation good. You're capable of putting your beliefs into action, although you prefer to

discuss them before doing so. Still, you believe that actions speak louder than words, so it's unlikely you'll be satisfied with all talk and no action for very long. Your timing tends to be especially good in virtually any kind of activity connected with athletics, writing, or graphic arts.

Mars Square Jupiter

This makes you want to experience everything you can. Your motivation is therefore extremely high. You can be very successful in your actions if you use common sense and do some planning beforehand; otherwise, you will have problems. Unfortunately, the tendency with this aspect is to leap into whatever interests you and then decide whether you're capable of handling it or not, which usually isn't a very constructive way to approach things. Your information tends to be scant and sometimes inaccurate; you tend to underestimate both the time and the difficulty involved in attaining your goals. In terms of timing, you're normally too quick to act (although if the square is in Cancer-Libra or Gemini-Pisces, you can waver until circumstances force you to move, and a combination of fear and procrastination can see you unprepared and on the losing end of the action). Regardless of whether you're too quick or too slow, your actions don't inspire confidence; your lack of preparedness can give people the impression that you're either desperate or just plain dumb! If you're too impulsive, slow down. If you're too slow, try to be more decisive. Either way, if you put in enough preliminary planning to be able to anticipate potential trouble spots, you'll do well. Trust to luck, though, and what luck you get is liable to be bad.

Mars Trine Jupiter

This shows excellent timing in virtually all activities, and you have a knack for taking chances and succeeding. Your energy level is high, but fairly easy to control; you're not as irrepressibly exuberant as with the conjunction or as reckless as the square. In fact, at times complacency with your lot can mar your motivation somewhat. This aspect seems to be especially beneficial in the creative area. It also often suggests that you'd be more apt to take up the cause of someone less fortunate than you than you would be to take action to benefit yourself alone. A lot of your action, therefore, is liable to involve helping or actively promoting someone else.

Mars Quincunx Jupiter

This tends to be a case of doing all the right things at the wrong time or perhaps for the wrong reasons. As a result, you don't always benefit from your actions, and when you fail, you don't always understand why. This can be a particularly annoying aspect in that sometimes your actions benefit others but actually work to your own detriment. There are sometimes misguided notions about competition—either you feel competition is bad and won't compete at all, or you'll compete only if you're certain you're going to win and are overly sensitive to failure (failure being anything less than first) You may consider yourself unlucky or you may feel that the game of life is somewhat "fixed" and only those with money and/or the right connections can get what they deserve (which you may feel they get by taking advantage of people like you).

Your energy level is erratic. You don't lack energy, but your self-discipline is such that you waste your energy on small tasks and then are tired out when big tasks need seeing to. Also, because this aspect gives a tremendous need for freedom, there's a tendency to want to do what you want do when you want to do it. This desire is so strong that you sometimes don't give enough thought to the potential impact your actions might have on your environment. Another tendency is to steep yourself in theories and knowledge, to study the experts in some area, and then do whatever they say without taking into account the fact that all theories, no matter how good, must be modified according to your circumstances and environment if they're to work for you.

You learn better from experiencing than you do from reading or listening. You therefore need to allow yourself to make mistakes, because unless you experience the results of your actions, you aren't motivated to change because you're not learning firsthand and can therefore detach yourself from the consequences of your actions. Look before you leap, take special pains to attune yourself to your environment and the people in it, and allow yourself to be less than perfect once in a while. Unlike some people, you learn from your mistakes, and what you learn helps you understand how to deal with your environment. Once you understand your environment, your timing improves tremendously.

Mars Opposition Jupiter

This is traditionally considered a lucky aspect, although I'd say it's lucky for the self-aware person. It can be a little less enjoyable for teenagers or for others who are not too sure who they are and what they want. In fact, I tell people with this aspect to be careful what they want and work for, as they're liable to get it. And if they later decide they don't want to pay the price, well . . . it can be rather tacky! Enthusiasm tends to be high, as does motivation. As long as you don't overexert yourself by taking on more than you can handle, you're virtually unbeatable. But if you try to play Superman or Superwoman you're liable to find yourself in some pretty messy situations. In other words, you have to know both your priorities and your limitations to make this aspect work for you. You're probably a physical person and for this reason should have some sort of physical outlet for your energy; otherwise, you can become impulsive or erratic. People with this aspect generally do rather well with sports and with things like modeling and dancing. More intellectual types may prefer things like psychology, media, or journalism.

What if there's no aspect between Mars and Jupiter? Well, that means opportunities are a bit fewer, and perhaps your ability to be in the right place at the right time isn't quite as good. In general, even people with squares or quincunxes have more opportunities, although they may not always handle their opportunities constructively. This doesn't doom you to plod along without hope of success; it just means you that you have to work a little harder to gain recognition. But remember, good timing is no substitute for talent. Everyone has talent; you have to find yours and develop it. When you do, opportunities will be waiting for you. And all it takes is one opportunity, constructively used, to put you on the path to success.

Chapter 6

Special Cases

There are certain conditions under which our interpretations of Mars have to be modified. These are when Mars is intercepted, retrograde, or under stress from a stellium. Let's look now at some of these special cases.

Mars Intercepted

Intercepted planets are presently a controversial issue and will remain so as long as people use different house systems. My opinion is that different house systems reflect different approaches to life—or perhaps different psychological layers of ourselves—and therefore which house systems "works" for you has to do not so much with scientific validity but rather with how you perceive your environmental priorities at different levels of your being.

For example, your intellectual self may view life best through an equal house lens, your material self through an equal house lens, your material self through a Placidus lens, your emotional self through a Koch lens, and your spiritual self through a Morinus, Campanus, or Porphyry lens. These different lenses show different views of life that somewhat alter priorities. And depending on which lens you use, different signs will be intercepted; only the angles will remain unchanged.

If you experiment with the various houses systems or lenses, you'll find that all have some striking fits—and all have some "clunkers." Eventually you'll choose the house system that works best overall for you, and undoubtedly that's the one you'll use. For me, that lens is presently Placidus. If for you it happens to be Equal House, that doesn't mean one of us is wrong; it just means we need different prescriptions for our glasses. It also means that if you use Equal House, you can safely skip this next section and move on to the material on Mars retrograde. If you're still with me, we'll look at the possible implications of an intercepted Mars.

So what happens with an intercepted Mars is that its potentials tend to be late-blooming because in your early years there was no adequate outlet for their development. Some people have trouble getting started; others start things well enough but run out of energy or enthusiasm before they finish them. Many people comment that they were held back from doing what they wanted to do for so long that by the time they were able to do it, they were too tired from the fighting to enjoy it or do it to the best of their abilities. Very often they talk about not being able to progress as fast as they would like. Sometimes they say they are overlooked by others and feel they have to push harder than others to get what they want or do what they want to do.

Here are some guidelines for interpreting intercepted Mars according to its sign and house position. As you read them, remember that while an intercepted Mars can indicate late-blooming potentials, those potentials are still there. They can and will manifest later in life.

Mars Intercepted in Aries

You want to lead or supervise others, but you can't for some reason. As a result, you tend to go your own way alone since you feel that if you can't lead, you'll at least see to it that you don't get stuck following for the rest of your life!

Sometimes this is an indication of wanting to do something connected with the involved house but letting fear of failure or rejection hold you back.

Mars Intercepted in Taurus

You apply yourself to your work or whatever you do in a methodical fashion but tend to get overlooked when promotion time comes. You want to acquire money or possessions but are somehow held back from doing so.

Mars Intercepted in Gemini

You have tremendous enthusiasm for reading and trying to learn, but something holds you back when it comes to using what you've learned. This could be your inherent tendency to scatter your energies or it could be some "outside" factor, maybe a declining need for people to work in your chosen field, maybe the economy, maybe having to get a full-time job in order to continue your schooling, et cetera. You have less ability than the typical Mars in Gemini person to adapt yourself to your circumstances, which may also hold you back.

Mars Intercepted in Cancer

Your intense emotions hold you back. You may, out of fear of being hurt, develop a strong wall of reserve. Or you may be hypersensitive and hard to deal with, thus making others leery of dealing with you.

Mars Intercepted in Leo

You work hard to attain possessions you want, but these are slow in coming. You may be inclined to take unwise risks because of your desire to make progress; if you start to feel too hemmed in, you tend to try resorting to ill-thought-out, last-resort attempts, which more often than not only tend to compound your problem. You're not, however, going to get rich quick or anything like that. Normally, you're a reliable and good worker who, like Mars in Taurus, tends get overlooked when the raises or promotions are being passed out.

Mars Intercepted in Virgo

You feel a need to criticize someone or something but are unable to do so. Or if you do get your chance, your comments aren't taken as seriously as you feel they should be. You want to act by yourself, on your own, but people hold you back or interfere with your plans. Your actions in general tend to be misunderstood.

Mars Intercepted in Libra

You feel a need to associate with others but are somehow held back. Either your community inhibits you or you are reluctant to get involved with others because you fear rejection. And while you're an industrious person, you're often taken for granted.

Mars intercepted in Scorpio

Your survival instincts are strong. As a result, you're among the least overlooked of the intercepted Mars people, as a rule. You feel a need to probe some subject or someone's depths (perhaps your own, perhaps someone else's), but something holds you back or makes it hard for you to get the information you need. People with this placement sometimes occupy themselves with obtaining metaphysical skills or knowledge to the exclusion of all else; this of course could hold them back in more mundane areas.

Mars Intercepted in Sagittarius

You feel a need to convince other people of the rightness of your ideas, but they are resistant. You display enthusiasm for your work, but for some reason others don't seem to be as interested in it or as excited about it as you are. Maybe you want to travel or emigrate but are for some reason are held back from doing so.

Mars Intercepted in Capricorn

You have the motivation to attain success and are nothing if not ambitious, but rewards for your efforts are slow in coming. Your self-restraint can sometimes hold you back from seeking recognition for your actions.

Mars Intercepted in Aquarius

You want to reform someone or something but your efforts meet with resistance. Your tendency to contradict established people and things may hold you back. Your plans tend to be harmonious with your goals, but realization of your goals may be slower than you'd like.

Mars Intercepted in Pisces
You may be inclined to search for some secret society that has the Ultimate Truth, and your search may hold you back in other areas. It could also be that the society you seek is so secret that you have tremendous difficulty finding it at all! Either an excessive fondness for (or an excessive dislike of) alcohol can hold you back. And your modesty can keep you from getting the recognition you deserve.

Mars Intercepted in the First House
There could be delays in getting started in life. You may have to force yourself to resist someone or something for your own good. Or you may be interested in a particular field or activity but unable to prove yourself in it for some reason.

Mars Intercepted in the Second House
There could be delays in achieving financial security or an inability to earn what you feel is an adequate living due to being held back. Or your values could be coldly materialistic and could hold you back in other areas.

Mars Intercepted in the Third House
There could be delays in education. Or thoughtlessness or feelings of hopelessness could hold you back.

Mars Intercepted in the Fourth House
There could be delays in forming family ties or in other areas somehow resulting from family ties. Weak will or moodiness could hold you back.

Mars Intercepted in the Fifth House
There could be delays in having children. Or you could be unable to meet all the demands placed upon you by your children, romantic interests, or social schedule without sacrificing your creativity in the process. Or it might be necessary to overcome an illness or handicap before you can realize your goals.

Mars Intercepted in the Sixth House
There may be delays in the work area. Or you may spend a lot of time thinking about leaving your job but be unable to actually quit for some reason. Or illness may hold you back.

Mars Intercepted in the Seventh House
There may be delays connected with marriage or someone or something may hold you back

from marrying your first choice; or you may marry someone you don't love because you've been pressured into doing so. It's possible you will want to divorce but be held back from doing so by someone or something. Or your own cynicism may hold you back from finding love.

Mars Intercepted in the Eighth House

There may be delays in getting money owed to you or delays in benefitting from partnerships or joint finances. Temper may be suppressed only to come out at inappropriate times, which can hold you back. Or "the powers that be" can be slow in hearing your prayers or intervening on your behalf.

Mars Intercepted in the Ninth House

There can be delays connected with the enjoyment of travel. Or you can concentrate your energies on expressing your philosophy of life to the almost total exclusion of all else. Or you may be unable to destroy certain prejudices or restrictive beliefs due to being held back by or meeting with resistance from others.

Mars Intercepted in the Tenth House

There can be delays connected with fulfilling your career goals or using your potential to the fullest. A physical or emotional illness or injury may hold you back. Or your physical development may be slow.

Mars Intercepted in the Eleventh House

There may be delays in connection with achieving your hopes. When angry you may be able to dish it out but not take it, or vice versa; this can hold you back. You may use inappropriate force sometimes and be inappropriately passive at other times.

Mars Intercepted in the Twelfth House

There may be delays in terms of extricating yourself from restrictive circumstances. Your actions may undermine your health in some way, thereby holding you back. Or your vitality or energy level may be weak. (I saw an interesting case: a woman had her home insulated with urea formaldehyde, and later began suffering allergy-like breathing difficulties that she felt were caused by the insulation. But because of the bad press urea formaldehyde has received, her home is now virtually unsaleable and she can't afford to move or to have the insulation removed. The woman's Mars is in Cancer intercepted in the twelfth house and quincunx Jupiter in the fourth.)

These aren't particularly encouraging interpretations. On the other hand, there's good news for those of you who have an intercepted Mars: you will outgrow the worst of the delays and inhibiting factors merely through living and experiencing, or at least you will if you do anything at all with your Mars potential. Exactly when you'll see a lessening of the delay factor depends on when your Mars will progress out of the intercepted sign. If you aren't familiar with progressions, you can make a rough approximation of when this is by doing the following: Subtract the degree placement of your Mars from 30. (For example, if Mars is intercepted at 7° Virgo: 30 -7 = 23.) Now multiply your answer by 1.5 (Mars takes about 1.5 years to go through the zodiac). So for our example, we'd come up with age thirty-four-and-a-half as the time when Mars breaks out of the intercept. This, of course, is a rough estimate; it could be a few months earlier or later. In any case, the thing to remember is that while you may be a late bloomer, you're by no means doomed to slogging your way through life without hope of reward.

Mars Retrograde

Most books say that when Mars is retrograde your energies are internalized or directed inwardly. There may be a tendency to go back over your actions frequently or to continually check up on yourself to make sure the efforts you're making are really representative of your true wants. People with Mars retrograde are frequently accused of being passive, disorganized, lazy, or hard to understand. While the latter may be true, the first two rarely are; it's just that you have a different way of using your energy than does the average person. In fact, I have come to believe that retrograde personal planets, including Mars, indicate some kind of exceptionality in terms of the functioning of that planet. There may be a giftedness, or there may be a handicap. Either way, you don't "do" what is symbolized the way everybody else does it.

In the case of Mars, the retrograde suggests that what motivates you is rather different from what motivates other people. What turns you on is not what turns other people on. And yes, you use your energy differently. This doesn't mean that you use it badly, but you can be conditioned to think it does. Society is, for the most part, inclined to cater to the average person's needs and behaviors. Mars retrograde says that you are not average, so what is provided does not meet your needs. This leaves you with two options: 1) you can try to play by other people's rules and norms, even though you may often feel like you're a square peg in a round hole and may also often find people accusing you of being inept, "not working hard enough," or whatever; or 2) you can look within for what is missing without and come up with your own way of doing things that allows you to maximize your potentials.

Either way, at some point in your life, Mars will most likely go direct by progression, and at that time, the way you do things will almost magically begin to jibe with the way other people do things. Meanwhile, your unflinching courage and determination should not be underestimated. What others see as mere muddling or dabbling can and will see you through an difficulty you find yourself in—and sometimes will provide a unique solution to what others see as problems.

That said, here are some guidelines for interpreting Mars retrograde by sign and house.

Mars Retrograde in Aries

Ambition and temper tend to be directed inward. As a result, you can be quite hard on yourself at times. You can, however, be truly inspired if you're in touch with your inner self, and this inspiration can help you realize your goals.

Mars Retrograde in Taurus

Endurance is directed inward. You have ability for creative work—particularly design and acting—and you also have ability for mediumship. Unlike Mars in Taurus direct, the product of your work is your chief reward; material gain is a secondary consideration. Other people may accuse you of lacking good taste because your concept of beauty is different from the average person's. This doesn't mean you actually lack good taste; it merely means there's a difference of opinion between you and the majority of society. Probably you shouldn't try to change to suit them—in fact you may not be able to even if you try.

Mars Retrograde in Gemini

You travel in an inner world rather than primarily in the outer world. Being sensitive, you're able to understand others almost instinctively; you may therefore be considered psychic or merely spooky, depending on those in your environment. You, however, would tend to consider yourself more logical than psychic as a rule. If you're in touch with your inner self, your actions can be truly inspired at times.

Mars Retrograde in Cancer

Impulsiveness is stifled, driven inward. You may feel you're an impulsive person by nature, but others only see your self-restraint. You're more concerned with gaining spiritual or psychological self-awareness than you are with understanding your environment. There can, for this reason, be a lot of "sampling" of various metaphysical disciplines, awareness movements, or psychological techniques. You feel an at-one-ness with others on an inner level and therefore are very reluctant to push, nag, or criticize them, as you feel that to hurt them you would hurt yourself as well. You may therefore be seen as more passive than you really are.

Mars Retrograde in Leo

Your sense of responsibility is directed inward. You feel your chief responsibility is to be true to yourself. You have ability for acting, dance, and other creative endeavors; you also have the ability to successfully deceive others if your sense of responsibility becomes perverted into a desire for self-glorification at all costs. If you try to adopt a sense of responsibility more like "everybody else's," a tendency to exaggerate your own importance may manifest as a defense mechanism. This happens when you please others, but you let yourself down in the process.

Mars Retrograde in Virgo

You adapt to what you feel are unpleasant changes in your environment by going inward for answers or advice on what to do. You tend to stifle your temper and are inclined to be much harder on yourself than on others. You have an intuitive understanding of others; your first impressions are nearly always correct. Others may attribute psychic powers to you, but you'd tend not to see yourself as a psychic person.

Mars Retrograde in Libra
Libra's desire to associate with others is turned inward; you want to get to know yourself better, be your own best friend. Enthusiasm is also inner-directed. You express only your higher feelings in love and friendship, and you keep your less noble sentiments to yourself and try to understand these and work them out. And your social and sexual preferences and needs may be different from those of your peers.

Mars Retrograde in Scorpio
Ambition is directed inward. Selfishness is stifled. The stirrings of your subconscious motivate you much more than any environmental factor. You're therefore a hard person to understand.

Mars Retrograde in Sagittarius
You travel on inner planes more than on outer ones. Frankness is turned inward; you're brutally honest with yourself. You have prophetic hunches, flashes of the future. You take them seriously, but if you discuss them with others you do so in an off-hand manner, perhaps almost jokingly.

Mars Retrograde in Capricorn
Ambition is turned inward so you're able to deeply immerse yourself in research, handle reams of facts and statistics, et cetera, with your prime concern being to prove something to yourself rather than to attain rewards. You also have a flair for succeeding in studies of the supernatural, in mediumship, or in stage magic. Whatever you do, people are apt to accuse you of wasting your talents; they're liable not to understand the purpose of your work.

Mars Retrograde in Aquarius
Teamwork involves your outer self and your inner self rather than you and others. Organization is turned inward, used for getting your own life in shape rather than cleaning up society. In the outer world, there's a search for soulmates or those known and loved in the past or at least those whose hearts are in tune with yours, rather than merely making do with whomever is handy to hang around with.

Mars Retrograde in Pisces
You tend to wait instead of act because you feel that if you're patient, whatever is wrong will right itself. If forced into a fast-paced, push-push environment, people with this placement often take up cigarette smoking; lighting up gives them an opportunity to wait a bit longer. Your

inner life is much more important to you than your outer life. No doubt some will misunderstand you and accuse you of being lazy. You're not.

Mars Retrograde in the First House

You feel a strong need for freedom. There's an inner restlessness, a constant searching for the most free environment. Your energy is spent on your own interests, which you constantly re-evaluate to make sure they truly interest you and serve your need for freedom. You're ambitious for gains, but your concept of gains is different from that of the average person; you seek inner gains more than material ones.

Mars Retrograde in the Second House

Your values regarding material goods, procreation, morals, contentment, money, et cetera, are somehow different from the average person's. You go over your financial situation again and again, quickly rejecting ways of making money that are not in keeping with your inner values. You double-check your motives before spending money for pleasure to make sure the pleasure will really be in proportion to the financial expense and the necessary expenditure of effort on your part. Your concept of ownership may be different from the average person's. Or you may strive to own something others feel is of little value.

Mars Retrograde in the Third House

Others may find you thoughtless or absent-minded because you tend to pay more attention to what's going on within than to what's going on without. You tend to repeat yourself when communicating or to often find yourself having to ask others to repeat what they said. This isn't so much due to inattention as due to a need to clarify things in your own mind, but others don't always understand this. You quickly grasp what's going on around you at an inner level but are slow to respond to it overtly. And at times you can be more preoccupied with yourself than others deem good. You tend to become impatient with yourself for not being able to make yourself clear to others or not being able to act on what they say without double-checking it.

Mars Retrograde in the Fourth House

If someone in your house comes down with flu or another infectious illness, you're liable to catch it. Mars retrograde here can indicate a recurring illness or depression caused at least in part by environmental conditions. You'll find yourself going back again and again to your childhood home or trying to search through the past for your roots. You tend to be slightly less secure than people with a direct Mars here. You're strongly ambitious in terms of wanting to make a home for yourself that expresses the real you—even if others disapprove of your plans. A parent may have influenced your sense of security for better or worse; possibly there was a traumatic event connected with a parent or with your home itself that occurred when you were too young to fully understand and deal with it successfully. There can be strife with those in your home or strife can be internalized and replaced with inner conflict.

Mars Retrograde in the Fifth House

Like Mars retrograde in the fourth, you're vulnerable to catching a virus or suffering recurring illness or depression, particularly if you have children or if you're very caught up in a romance or social life. You'll find yourself going back over your good times again and again and trying to recreate them. If you have children, one may have to repeat grades or subjects in school or often have to redo work. You tend to take the same risks more than once. Your fondness for children at a certain age may cause you to have several so you can keep repeating the joys of that age. There's sometimes a tendency to rehash arguments again and again—especially with lovers or children.

Mars Retrograde in the Sixth House

Others may describe you as nervous or weak, or accuse you of misusing your energies. Sometimes prescriptions have to be renewed because the first batch didn't totally clear up the problem; sometimes surgery has to be repeated for one reason or another. Health problems, in any case, have a way of recurring, especially headaches. Work may be repetitive or frequently have to be redone. Or you could quit a job only to return to it after exploring other options. If accident-prone, accidents would tend to be of the same kind or to always involve the same part of the body. Contrary to popular belief, you're not weak. You do, however, tend to be more at ease in your inner world than in your outer environment and you tend to spend more time strengthening your inner self than your outer body.

Mars Retrograde in the Seventh House

You're vulnerable to catching illnesses and infections, particularly from your mate or those who are especially close to you. You tend to want to repeat the joys of your relationships—second honeymoons, renewals of marriage vows, and returns to where you first met are common. Sometimes there's a delay in finding marital fulfillment because you put fulfillment of your inner potentials first. There may also be delays in terms of consulting with doctors since there's an inclination to try to heal yourself from within or to see illness as being caused by your state of mind rather than being tangible and organic. Lawsuits are sometimes ill-timed; you may wait until the last possible minute to take action. Energy tends to be directed inward rather than into partnerships; you want to form a partnership with your inner self before opening your arms to others. This can be a source of difficulty in partnerships; while you see yourself as striving to give the other person the best—most together—self you have to offer, the other person may misinterpret your attempts to get your act together as selfish or narcissistic. You may choose an outwardly-aggressive partner so you can concentrate on inner development and let him or her handle the more mundane details of living.

Mars Retrograde in the Eighth House

If another damages your life in any way, your tendency is to give this damage back in kind, thus setting up a cycle of hurt begetting hurt. Others may accuse you of being brutal; you see this merely as being fair. Conversely, if you receive help or kindness, you'll go back again and again to repay the debt. Sometimes there's almost a compulsive need to organize things that others value or to organize people of like minds to work for a common cause; this work may have you repeating your search for what you collect in a great many places. There may be recurring fevers or headaches. Financial matters involving others tend to be checked and rechecked before being accepted. Forcefulness turns inward; you force yourself to do what others find distasteful.

Mars Retrograde in the Ninth House

Your concept of wealth, luck, or progress may be different from the average person's. You tend to hold back your opinions and keep your philosophy of life to yourself. In-laws pop in and out of your life, even if you divorce. You may attract foreigners from one specific country. You may have an interest in a particular facet or religion, philosophy, or education that you keep going back to no matter how many times you drop it. Or you could drop certain religious or metaphysical views at one point only to go back and embrace them even more strongly later on.

Mars Retrograde in the Tenth House

Others can mistake your searching for instability. You may build up a business only to find you have to rebuild it because of relocation, illness, or change of public opinion. Or you may be someone who rebuilds failing businesses as a hobby. Jealousy can be a recurring problem in the career area. One success leads to another. Or you can have several successes in the same field. Self-reliance tends to be more important to you than public recognition or approval.

Mars Retrograde in the Eleventh House

Your energy level seems to fluctuate because today's goal may not be tomorrow's, but three days from now it may again be your goal. It takes you a bit longer to make changes in your circle of acquaintances because you must feel the need for change on an inner level before you take any action to find more satisfying companionship. You're therefore inclined to have more non-productive relationships (according to others, at least) than the average person because unless you have a gut-level conviction that someone doesn't belong in your life, you won't take action to remove the person. Friendships are sometimes broken off only to be restored later; you can lose touch with people only to have them return later. Others tend to see you as passive in terms of attaining your hopes because they don't realize how hard you work. Contrary to what you might expect, when Mars is here the realization of your hopes depends more on your efforts than on any group involvement or professional association. Friends may see you as taking a passive attitude toward them; they may not realize how much effort you put into your friendships.

Mars Retrograde in the Twelfth House

You seem to repeat the same mistakes because you have to learn the hard way. Curiosity is directed inward; you want to know what makes you tick. You may at times repress your feelings when you shouldn't; this can hamper your growth potential. When this happens, resentment can develop and cause recurring problems.

A retrograde Mars is neither good nor bad; it's merely different from the norm. If you have Mars retrograde, "To thine own self be true" should be your motto. The more you try to succeed by doing what everybody else does or feels you should do, the greater your likelihood of failure.

For those of you who are uncomfortable with retrograde Mars, it may make you rest easier to know that this, too, tends be outgrown by progression, provided you don't just ignore inner motivation.

Mars Under Stress from a Stellium

Any stress aspect to Mars makes energy harder to direct or makes it harder to assert yourself in a constructive manner. When a stellium of planets gangs up on Mars, it gives you an inclination to switch rather than fight. In other words, you may begin to act like your Mars was really in the stellium sign. But of course it isn't. Your motivations are somehow different. So trying to be something you're not in terms of energy can lead to frustration and/or inner conflict. The inner conflict no doubt comes as no surprise to you. I mention this factor merely to explain why you might identify with Mars in your stellium sign instead of (or in addition to) Mars where it actually is. Sometimes, there's also a sort of subconscious synthesis; for example if you have Mars in Pisces and a stellium in Virgo you may identify with Mars in Gemini or Sagittarius in addition to (or instead of) your own placement.

This tendency tends to be most marked when you have the Sun or Moon in a stellium that makes stress aspects to Mars. It's most difficult for the woman who has the Sun in a stellium square Mars, and for the man who has the Moon in a stellium square Mars. In these cases there tends to be dissatisfaction with Mars the way it is. As a result, there are attempts to act more in accordance with the stellium sign than with the Mars sign, and to feel dissatisfaction with your accomplishments.

The conflict between Mars and the stellium tends to be most severe in the romantic area. With the stellium emphasis, you tend to identify with those members of the opposite sex who typify the characteristics symbolized by the stellium because you're comfortable with these traits. At the same time you're attracted to and want certain characteristics connected with your Mars even though certain other characteristics annoy you. So women who have the Sun in a stellium square Mars, and men who have the Moon in a stellium square Mars, may not have found the Mars-based descriptions in the love-life chapter totally accurate. You may instead identify with a description of Mars in your Sun or Moon sign or a synthesized Mars position based on a mid-point sign. Or you may feel that each of these positions has its good points but also gets on your nerves. Yyou'll have a harder time than the average when it comes to finding Ms. or Mr. Right.

Chapter 7

Progressed Mars

As we grow, physically and psychologically, the things that motivate us—and irritate us—change. These changes in motivation, in energy level, and in what we will and will not put up with are shown by the position of progressed Mars.

There are many different systems of progression (or directions, as some of these systems are called). I won't get into the how-tos of these many systems here, as that's not the purpose of this book. And I don't really like praising one system over another, as I feel that the reason there are so many systems is that there are many different types of people. We all "unfold" at different rates, we all have different environments that condition us differently, and I think that this is why, as with houses, what works very well for one astrologer may not work so well for another. Arguing the merits of one system over another is like arguing over whether blue is a better color than red.

As I mentioned in the previous chapter, if you were born with a retrograde Mars, it will no doubt go direct by progression. And if you were born with a direct Mars, it can go retrograde by progression. The points where Mars changes direction—called stations—are also very important and generally indicate turning-point years. When Mars goes from retrograde to direct, there is normally a change of motivation, a change of direction, and a change of behavior. You become more direct and more involved in the outer world. Energy and passion may both seem to increase. You may begin acting on plans that have been gestating for years. Or you may cut something out of your life. Specifics would of course depend on the sign, house, and aspect, but one thing's for certain—your actions will be noticed and suddenly you will be accused of doing the "right" thing.

If, on the other hand, Mars goes retrograde, there is often a walking away from something. You may become quite assertive about not wanting to put up with this any more, or you may simply walk quietly away. Sometimes you are motivated to go back and revisit past actions or activities. Sometimes you go back to finish something you started and never finished. Note that if Mars goes retrograde after your birth, it frequently stays retrograde for the remainder of your life. The same is true if Mars goes direct after your birth. Either way, the stations denote a major change of motivation and generally a major lifestyle change of some sort.

What you have to remember with progressions and directions is that they're symbolic measurements. They're based on some formula such as one day's planetary movement equals one year of life-experience, or the average twenty-four-hour motion of the Sun (59'8") equals one year of life-experience. The only way you can decide which of the many symbolisms best fits your life circumstances is to try the different systems.

Remember, too, when looking at the following material, that you never lose the symbolism of your natal position, although you become more of what the progressed sign position symbolizes. For example, I have natal Mars in Pisces, which is a pretty laid-back position. At age sixteen, my progressed Mars went into Aries. This didn't turn me into a trailblazer overnight, but it did bring a turning point; where earlier on I'd felt helpless to change what I didn't like in my environment and coped by withdrawing and daydreaming, I suddenly started speaking up—and fighting back! But I still was not, and never will be, a Mars in Aries person. My interests lean a lot more toward the occult, the arts, and passive stuff, than they do to sports and highly physical stuff. My energy level, while higher in terms of stamina, certainly is much lower than that of a person born with Mars in Aries. And while I'm quicker to speak up about things, I retain the natal Mars in Pisces tendency to brood before acting. Remember, everything's relative.

Progressed Mars in the Signs

Progressions represent a subtle evolution—not a total makeover. With this in mind, let's look at the various progressed Mars positions.

Progressed Mars in Aries

After a period of going with the flow, waiting, seeing, and putting up, you become fed up and are now in more of a fighting mood. You become quicker to take action, quicker to anger. Your patience, and sometimes your tolerance, decreases. You suffer fools less gladly. At the turning point of this progression (around the time Mars hits 0° Aries), you can be overtaken by a bout of blind enthusiasm that can, if you're not careful, get you into trouble. Throughout this period, there's an increased danger of injuries to the head, face, et cetera; however, before you panic, remember that for something serious to happen along these lines, your natal chart must show the potential for it to happen and transits must indicate that there's an opportunity for this to happen.

Progressed Mars in Taurus
After a period of starting things with enthusiasm, but being inclined to drop them when they lose their newness and become routine, you now have an increased capacity for hard work and an increased inclination to see things through to the bitter end. Determination increases, as does obstinacy. But where you previously were inclined to rush in and push, you now prefer to make your moves slowly and steadily.

Progressed Mars in Gemini
After a period of being housebound or quite settled in one environment or routine, you now become more mobile. After somehow putting all your eggs in one basket, or focusing very closely on one specific thing, you feel you need a change. So at the turning-point (around the time Mars hits 0° Gemini) there's a strong tendency to scatter your energies; some of this lack of focus can permeate the whole cycle. Gemini is connected with learning, so you'd expect yourself to be more highly motivated to learn in this period. It's true that there is an increased attraction to learning, but what draws you are spontaneous sorts of learning experiences—not classrooms and cramming for exams. Disciplined study interests you less than before; you don't want to put all your energies into developing your mind. So, at this time, mini-courses and learning-by-doing experiences are more to your liking than heavy research or degree programs.

Progressed Mars in Cancer
After a period when "keep it light" was your motto, your emotions now become more intense. Your self-control lessens. Sometimes your sense of direction in life changes at the turning-point period; sometimes your emotions temporarily obscure your sense of direction. You're motivated to find security now. What constitutes security and how you go about getting it must be determined by your natal chart and your progressions and transits at the time.

Progressed Mars in Leo
After going through a very security-conscious and somewhat conforming phase, you now become more creative as well as more inclined to take risks. As a rule, you're more inclined than usual to toot your own horn during this period. Humility decreases—particularly if you're successful in your risk-taking.

Progressed Mars in Virgo
After a cycle of high creativity and big plans, your imagination moves into the background, decreasing somewhat. Your interests might shift from the purely creative to the more technical, from the forest to the trees. You become more concerned with tidiness and order.

Progressed Mars in Libra
After a cycle when work or duties may have kept you too busy to do much strictly for fun, you now feel an increased desire to socialize. Trouble is, because you've been out of the scene for a

while, you may feel a bit out of your element. Suddenly, being liked is a lot more important to you. So your independence may become a bit shaky, and you may gear your own actions to other people's feelings to a greater extent than usual. Your affections are stimulated to a greater extent as well. Enthusiasm takes an upswing.

Progressed Mars in Scorpio

After a cycle where you were primarily motivated by other people—and sometimes perhaps not motivated in directions that were in your own best interests—you feel it's time to look out for number one a bit more. So your survival instincts become a bit stronger. This can be good, but it can also lead to a tendency to waste energy in defensive behavior or in unnecessary self-protection ploys—if your natal chart inclines you toward that sort of behavior. It may become more difficult to relax; tension can be more of a problem than it usually is.

Progressed Mars in Sagittarius

Your need to convince others that your own ideas are best continues to grow, but you become less defensive and less wary, as a rule. You are more attracted to contests, more willing to take risks, more impulsive. Your powers of endurance may decrease somewhat.

Progressed Mars in Capricorn

After a cycle of going pretty much with the flow and perhaps trusting to luck, fate, or "good aspects" a bit more than was wise at times, you now feel it's time to see some results for your actions. You therefore become more ambitious, more serious. You have less time than before to spend on jokes and fooling around; your sense of humor may be less in evidence or may take on a cynical edge. Your motivation to succeed can lead to success, but as you succeed there's an increased tendency to over-estimate yourself, which can cause problems if your natal chart inclines you toward ego problems.

Progressed Mars in Aquarius

Your concern with success in the here-and-now develops into concern with success in the future as well, so you continue to act with increasing deliberation. But where before you were willing to conform or at least play the game, now you're more in the mood to contradict others or play devil's advocate. You become more detached from others and what they think, more determined to set your own course even if others don't approve.

Progressed Mars in Pisces

After a busy cycle of taking deliberate action, you're a little bit tired so you become more inclined to wait instead of act. Your energy level may, in your opinion, become less adequate than before. It may fluctuate more; you may tire more easily or need more sleep than before. In short, you become more relaxed, less driven or pushed by your energies. This can be a plus, but it can also mean that you become less assertive and less disciplined. In other words, in some cases your self-control decreases and an apathetic or "what the heck, nobody else does it" attitude can manifest. Also, when Mars goes into Pisces, you become less interested in the outer world and more interested in an inner one. As a result, there's an increased interest in the occult that can encourage you to join a secret society or metaphysical group, or take yoga lessons, for example.

Progressed Mars in the First House

After a period of enforced or self-imposed isolation, separation, or soul-searching, you emerge to take a more active role in your environment. Assertiveness increases; sometimes, when progressed Mars contacts your Ascendant, you go through an aggressive phase, perhaps in an attempt to overcompensate for all the times when you for one reason or another couldn't or didn't assert yourself. During this period you may take various forms of action to improve your health or appearance; you may get involved in a sport, exercise program, a behavior modification program, or even have cosmetic or corrective surgery. Usually your sensitivity to heat is increased in this period. You sweat more and possibly have more trouble with skin eruptions as a result. And when you get sick, you're likely to run a slightly higher temperature than before.

Progressed Mars in the Second House

Having worked on yourself and your lifestyle to some extent, you're now ready to become more enterprising in terms of earning resources for yourself. If your age permits, you might take some sort of work-related action; if not, you might redefine or determine your values or put them into action. You could get involved with a Mars-type business or activity. There might be gains (or losses, depending on your overall chart) connected with life, machinery, psychology, or other Mars-related activities or things.

Progressed Mars in the Third House

In the process of acquiring things, money, or values, you undoubtedly learned a thing or two and got some new ideas. Now you think about taking action on these. (Whether or not you actually do act on them would depend on what's being activated in your chart at the time). In addition, you may take action to improve your neighborhood or see a lot of activity involving siblings, cousins, or other third-house people.

Progressed Mars in the Fourth House

After a period when acquiring ideas or feedback was the main theme, your material acquisitiveness increases. This may be a little or a lot, depending on your overall chart pattern. Usually the things you want in this period are relatively practical or security enhancing; they may be things for your home or a home itself or land or something related. Sometimes it becomes necessary to acquire things; household appliances wear out or malfunction with alarming frequency in this period. In extreme cases, there can be damage to your home, and also in the extreme category, someone—including you—could be injured in your home. For these things to happen, however, there would have to be some indication in your natal chart that this sort of event is possible.

Progressed Mars in the Fifth House

Having acquired a home or things for it, you now may take steps to see that what you've gotten will be treated appropriately. So you may become stricter with your kids—if you have kids—or more demanding of your mate, lover, or anyone else who spends a lot of time under your roof. Since you've been working hard, you may feel you need a break and may get involved in some sort of hobby. This might be a sport or course, and in some cases would be something that takes a fair amount of mental energy and in some way allows you to express yourself. It's also possible, for one reason or another, that you won't get directly involved in sports yourself, but rather a child or a lover gets involved in something of this nature and you spend a lot of time watching games or performances or taking your child to practice or lessons.

Progressed Mars in the Sixth House

After a period when social or recreational pursuits may have used up a lot of your energy, you become more seriously or sensibly industrious. What you buy at this time will have to be serviceable. The turning point of this cycle, could prompt you to, for example, do an overhaul on your wardrobe and buy new clothes; what you tend to buy is active wear and sports clothes rather than formal wear and delicate fabrics. You may become motivated to address your weight or your diet. An interesting phenomena that sometimes occurs is a craving for hot or spicy foods; this fondness may continue throughout the cycle.

Progressed Mars in the Seventh House

After a cycle of hard work, you're willing to let others take the initiative. This often tends to be a problematic cycle because of a tendency to give away your energy and motivation and let someone else determine what's to be done with it. The outcome of this can be relationship problems of one sort or another as you come to see that the person you gave your energy to isn't interested in going in the direction you want to go in. In extreme cases, your dissatisfaction can lead to a lawsuit or divorce.

Progressed Mars in the Eighth House

After a period of listening and accepting, you're now in the mood to devote your energies to some sort of research. Often this is inner-directed research into your own psyche, through psychology or some branch of the occult, although sometimes energies are expressed more overtly through an involvement with heavy industry, manual labor, or another form of Mars-related work. Your energies in general are intensified. Your sex drive in particular often increases.

Progressed Mars in the Ninth House

After a period of looking within and being fairly conservative, you might go through a period of dissatisfaction with "where you are" on one level or another. This might mean that you go on a journey to find a more compatible place to live or a more satisfying philosophy to live with, or it might mean that you return to school in the hopes of learning something that will put you in a better space. Occasionally, people become involved in political activism. Once in a while there are injuries in the course of travel.

Progressed Mars in the Tenth House
After going through a searching, idealistic, and somewhat future-oriented phase, you prefer to use your energies in here-and-now life. Your ambition increases. You may start a new career activity—particularly around the time Mars hits the Midheaven. Politics, psychology, the police, or other types of Mars people and things may figure into your life more heavily than usual.

Progressed Mars in the Eleventh House
Strong desires connected with the future surface. You can find yourself or those around you espousing revolutionary sentiments, and friends can encourage you to take risks, or vice versa.

Progressed Mars in the Twelfth House
After concentrating heavily on the future and the present in relation to the future, now you're motivated to reflect on the past. Sometimes this reflection is purely voluntary; sometimes it's kind of forced on you by a painful experience. Or perhaps an injury or illness at one point leaves you with a lot of time to think and not much capacity to get up and go. The results of your reflection depend on how you perceive your past. If you believe you've generally acted to the best of your abilities, this may be a time when you seek a greater understanding of why things went wrong, why they succeeded, and how your actions contributed in each case. If you can't accept your own limitations or the consequences of your actions, you may become angry, act in counterproductive ways, and as a result become your own worst enemy. In extreme cases, some people experience self-hatred because of past actions. Should this happen, there's a danger of self-destructive impulses of some sort surfacing.

Progressed Mars Conjunct Natal Sun
You want to be in the limelight. Your activity level increases. Often you channel your energies into sports or other types of Mars activities.

Progressed Mars Conjunct Natal Moon
Your willpower increases, which is a good thing. This is often an irritating time when a lot of little annoyances crop up. You may find yourself having to deal with aggressive people, and may respond by becoming more aggressive yourself. Patience wears thin as hassles mount; as a result, you may become more impulsive, so the danger of accidents increases somewhat. You may buy a home or things for your home; you may sell your home or things in it.

Progressed Mars Conjunct Natal Mercury
You have a flair for quick repartee in this period. You might be actively investigating something or taking steps to learn a skill. You might make an important lifestyle-related decision.

Progressed Mars Conjunct Natal Venus
Your sensitivity increases, and you could be involved in a big romance and possibly even marry. You could start a new business activity, and there might be financial gains.

Progressed Mars Conjunct Natal Mars
This aspect merely emphasizes the message of natal Mars.

Progressed Mars Conjunct Natal Jupiter
You're probably concentrating on a goal and using your energy to accomplish what you have set out to do. Things that might stand in your way at this time are religious fanaticism, prejudice, or a "might makes right" attitude. These could coming from within you—if you're inclined that way—or come at you from others.

Progressed Mars Conjunct Natal Saturn
Right now you're being driven by your ambitions. This aspect sometimes brings a job change or involvement in heavy industry. Relationships often have a harshness to them during this period, in part because you're so determined to get your own way that you fail to see the impact your actions have on others. For this reason, people sometimes experience a literal or figurative hurt at this time.

Progressed Mars Conjunct Natal Uranus
You become more obstinate, more impulsive, and sometimes more careless. Sudden accidents and endings—both literal and figurative—often occur at this time as a result. These may happen "to" you or they may affect you less directly. Sometimes people get involved with guns at this time.

Progressed Mars Conjunct Natal Neptune
Your energy level decreases now—sometimes to the point where you become apathetic or self-indulgent. There can be problems with drugs at this time—not just abuse or addiction, but things like allergic reactions or intolerances as well. There can be dealings with institutions and/or with religious or metaphysical people.

Progressed Mars Conjunct Natal Pluto
You want what you want and you'll use force to get it if you have to. What you want is to eliminate some old, outgrown, or counterproductive condition. Will this be easy? Probably not. Will there be fights? You bet! Endings and injury are both potential themes; you can hurt others, get hurt yourself, or a bit of both. The injury may be literal or figurative. This progression seems to show up frequently when people enlist in the service (or when drafted).

Progressed Mars Conjunct Natal Ascendant
You're looking for gains and advancement, and you're willing to use force. Self-centeredness or a "me-first" attitude can surface. Usually there's a lot of excitement in your environment for better or worse.

Progressed Mars Conjunct Natal Midheaven
You are concentrating on a goal. Occasionally there's a conflict with a parent or authority figure at this time.

Progressed Mars Conjunct Natal North Node
You may not be as adaptable as usual. You're taking some sort of definite physical action in order to grow, but to paraphrase the old Frank Sinatra song, you're doing it your way. So if you're trying to gain popularity, it's a very specific popularity rather than an "all-around best-liked person" award, and while you may win friends for what you do, chances are you may also attract some disapproval. Often you gain new friends but receive disapproval from old ones. It's possible you could be involved in some sort of competition in hopes of gaining recognition.

Progressed Mars Sextile Natal Sun
Interest in romance—and sex—takes an upswing. Sometimes there's a pregnancy, sometimes an increased involvement with children. There's an opportunity to become involved in sports or fitness activities.

Progressed Mars Sextile Natal Moon
You become more forceful, more dynamic in the expression of your emotions. You have an opportunity to lay your emotional cards on the table, to get things off your chest. An introduction to new cuisines is possible, often including spicy foods.

Progressed Mars Sextile Natal Mercury
You become increasingly fond of discussing and debating. You might have an opportunity to learn a new skill or craft or to increase your skill in one you've already learned.

Progressed Mars Sextile Natal Venus
Your sensuality increases. There might be a new romance or some sort of partnership activity at this time. Or there could be a work opportunity.

Progressed Mars Sextile Natal Mars
This can bring an opportunity for something connected with pursuing an interest in science, psychology, engineering or another Mars-related activity, or to become involved with uniformed military or other Mars-type people. You could do something that involves using strategy.

Progressed Mars Sextile Natal Jupiter

Here's an opportunity to constructively direct your willpower. You could become involved with sports, travel, manufacturing, organizing, teaching, or something similar.

Progressed Mars Sextile Natal Saturn

Your powers of endurance are increased. You could do something that involves tools or hardware. Also, you might have an opportunity to get some dental work done before you have to get it done. Elective dental work is also possible.

Progressed Mars Sextile Natal Uranus

You could take action connected with or involving friends, or get involved in astrology or occult studies. There could be some sort of extraordinary opportunity connected with the involved houses.

Progressed Mars Sextile Natal Neptune

You have an opportunity to refine, sublimate, or transcend your feelings or to develop a more positive emotional outlook. You could take action in connection with an artistic or creative interest.

Progressed Mars Sextile Natal Pluto

You have an extraordinary amount of energy and stamina at this time, so you can be quite forceful if you choose to be. You could become involved with police, factory workers, athletes, et cetera.

Progressed Mars Sextile Natal Ascendant

Your forcefulness and willpower increase without being obnoxious. Rather, you're trying to get your own way through actively cooperating with someone or in something. An important relationship could be formed in this period. There could also be activity connected with physical self-improvement.

Progressed Mars Sextile Natal Midheaven

You're ready for action, and you should see an increased activity level at home, at work, or both. You might become involved in a home improvement project, and there could be a work-related opportunity.

Progressed Mars Sextile Natal North Node

You want to cooperate with others and will do your best to do so. You might take action to benefit from a trend within your environment, as well as becoming increasingly interested in sports, fitness activities or other Mars activities as a means of growth.

Progressed Mars Square Natal Sun
Obstinacy increases, and frustration can lead to something less than good judgment; in this case, an accident might occur. There might be some involvement with military, police, or other Mars types, which could be voluntary or involuntary.

Progressed Mars Square Natal Moon
Your emotions are stronger than usual; you're more excitable, more irritable. This could lead to some people problems or adverse attention. Emotions getting the better of you could also lead to some problems with your nerves, muscles, or digestion.

Progressed Mars Square Natal Mercury
You have an increased tendency to exaggerate at this time. You might have to deal with occupational hazards or less-than-pleasant environmental conditions.

Progressed Mars Square Natal Venus
There is an increased tendency to exaggerate, and with it a decrease in refinement. You may feel that others aren't considering your emotional needs, and this may lead to scenes or an emotional "tit for tat." ("If they don't care about me, why should I worry about how they feel.") Obviously this isn't the niftiest time for romantic involvement. There may be difficulties with artistic or creative projects as well.

Progressed Mars Square Natal Mars
Depending on which progression system you use, this would occur at an advanced age—around age 90 or even later. It might coincide with wanting sex but being unable to have it, or with feverish illnesses, accidents, or burns.

Progressed Mars Square Natal Jupiter
You're rebelling against rules, and there's a danger that your behavior will be somewhat less than ethical. There could be problems with higher education, religion, travel, et cetera.

Progressed Mars Square Natal Saturn
If you've been harboring anger or resentment for a lengthy period, watch out! You could concentrate your energies on "getting" someone or "showing" someone a thing or two. At the very least, you're demonstrating self-interest in whatever you do. At worst, this could bring out a real mean streak—even a violent streak if you or someone close to you is inclined this way. A less severe manifestation would indicate problems with your teeth.

Progressed Mars Square Natal Uranus
You're in a more than usually argumentative mood and are less inclined to be generous and to give the benefit of the doubt. You might even become downright selfish for a bit. Your life

could be disrupted by some sort of circumstances beyond your control; this disruption is apt to be connected with the involved houses.

Progressed Mars Square Natal Neptune

You may be consciously trying to direct your energies along certain lines but for one reason or another may be unable to take appropriate action. You're dissatisfied with yourself and irritable because of this; you're often mad or fed up without really knowing why. Secrets or hidden things may have to be uncovered and dealt with; literal or figurative "hidden enemies" may have to be dealt with. In extreme cases, frustration can lead to criminal activity, substance abuse, or some other type of escapist behavior—even suicide. (This is assuming of course that the natal chart shows potential for this sort of behavior; if it doesn't, these extremes won't manifest.)

Progressed Mars Square Natal Pluto

You're angry and you're not gonna take it any more! You're going to get what you want, no matter what it costs in terms of time, money, or energy. You can be ruthless yourself, or ruthlessness can come at you from others. Normally this is a psychological ruthlessness, but in extreme cases, if your chart allows for the possibility, there can be physical violence, even rape. There can be counterproductive sexual activity, or there can be sexual problems. There can also be conflict over the use of jointly held finances or other shared resources.

Progressed Mars Square Natal Ascendant

You could be dealing with quarrelsome people or you could be in a quarrelsome mood yourself. Either way, you can expect a certain amount of conflict in this period. Sometimes sexual desires intensify.

Progressed Mars Square Natal Midheaven

You're in an excitable mood. The reason may be work problems, loss of a job, domestic problems, forced relocation, or another lifestyle change that's not entirely of your choosing.

Progressed Mars Square Natal Nodes

Discord is the theme. You feel unable to cooperate with what's going on around you. You don't respect tradition or the norm, but you don't really have anything to offer in terms of a better way—or at least you have nothing to offer that meets with other people's approval. Your disdain for or frustration with your environment may lead to anti-social attitudes or behavior.

Progressed Mars Trine Natal Sun

You might find yourself accepting a leadership position. Or you might start a self-improvement activity. You could enjoy sports, fitness programs, et cetera, to a greater extent than usual.

Progressed Mars Trine Natal Moon

You use a lot of energy to start something. Your ambitions—and what you start—might involve the work or domestic areas. This progression often coincides with engagement or marriage.

Progressed Mars Trine Natal Mercury

You're in an ambitious mood and working to achieve success. An important agreement can be reached, or an important contract can be signed. You might take action to improve your health, start an exercise program, sign up for a course, et cetera.

Progressed Mars Trine Natal Venus

You might start an important sexual/romantic relationship, which may or may not lead to marriage. Financially, you're in an ambitious mood. Sometimes this is a period when for some reason you receive many nice gifts. Or you could receive a big financial boost.

Progressed Mars Trine Natal Mars

In most progression systems, this aspect would occur at such an advanced age that it would not be relevant.

Progressed Mars Trine Natal Jupiter

You might reach an important agreement, sign a contract, or succeed with a legal matter. Generally, people have an increased concern with honor, ethics, et cetera during this period. Earning a university degree is feasible.

Progressed Mars Trine Natal Saturn

You might use your energy to overcome some sort of difficulty or obstacle; the more energy you use, the more difficulties you'll overcome. You also invest your energy into some sort of practical work, either with a group or on your own.

Progressed Mars Trine Natal Uranus

You might achieve something unusual, possibly connected with technological innovation, electronics, astrology, psychology, or science. In any case, you're probably doing something that involves using an extraordinary amount of energy and putting forth a tremendous amount of effort. You could deal forcefully with people or merely be dealing with forceful people. Friends may play a major role, for better or worse, in what occurs during this time.

Progressed Mars Trine Natal Neptune

Inspiration may strike in some way. You can make progress in connection with secret or hidden goals or plans. Sometimes something happens behind the scenes that is to your benefit; as a result you may progress without being aware of it. You can benefit in connection with non-profit organizations, hospitals, creative endeavors, et cetera.

Progressed Mars Trine Natal Pluto

You're putting forth extra effort to achieve success. You might get involved in an athletic activity, or there could be progress connected with physical or psychological development of some sort.

Progressed Mars Trine Natal Ascendant

You're driving yourself—and possibly others—harder than usual. You're more passionate than usual.

Progressed Mars Trine Natal Midheaven

Your determination and resolve increase. You could attain a life goal by taking definitive action, and there could be a promotion or another career-related event.

Progressed Mars Trine Natal North Node

This suggests a growth-related achievement, possibly involving other people.

Progressed Mars Opposition Natal Sun

Strain is likely, possibly a great deal, and upsets and/or injuries are possible. In extreme cases, this aspect can coincide with a heart attack—again, assuming the potential is present in the natal chart.

Progressed Mars Opposition Natal Moon

You're less tolerant and more headstrong than usual. You experience incompatibility of some sort with other people or your environment; inner conflict is also likely. There can be sexual, romantic, or marital difficulties, as well as problems in dealing with your parents and with women in general.

Progressed Mars Opposition Natal Mercury

Disputes, debates, and arguments occur with greater than usual frequency. Car accidents occasionally occur. Resistance to illness can be a little low, and if you have allergies or respiratory problems they may give you more trouble than usual during this period.

Progressed Mars Opposition Natal Venus

You're less generous—perhaps out of necessity, perhaps not. Hassles connected with money are common, as are marital problems. Occasionally this coincides with a romantic relationship that isn't in your best interest.

Progressed Mars Opposition Natal Mars

In most progression systems, this aspect would occur at such an advanced age that it would not be relevant.

Progressed Mars Opposition Natal Jupiter

You might be involved in important negotiations or in trying to resolve a conflict. Or you might refuse to cooperate and see a lawsuit as the result. Your sense of self-importance can be increased; sometimes this is a healthy thing, but other times it's the sort of pride that goes before a fall.

Progressed Mars Opposition Natal Saturn

This is often a period when you feel—or actually are—rather helpless. Circumstances beyond your control tend to play a greater-than-usual role, and almost everything you do seems to meet with resistance. In other words, your strength is tested. Sex drive decreases; tiredness increases. Sometimes impotence or other sexual problems occur. Someone close to you may become ill, injured, or otherwise unable to take care of themselves; as a result, added responsibility can fall on your shoulders. Your own health can suffer.

Progressed Mars Opposition Natal Uranus

Your nerves are very definitely tested. You're fed up; you want to overthrow or get rid of some existing structure or situation. Conflict results. This aspect very occasionally coincides with births—particularly births where the timing or circumstances are difficult.

Progressed Mars Opposition Natal Neptune

There's some sort of subtle or hidden danger or problem. This can lead to a failure caused by lack of planning or by lack of awareness. Occasionally water damage to property can occur during this period.

Progressed Mars Opposition Natal Pluto

You're going to go ahead and do what you want to do regardless of what other people say. You can be ruthless—even brutal. Or ruthlessness or brutality can be directed at you from competitors or those who are opposed to your plans. You may feel a need to rechannel some of your sexual energies or to transform your sex life. Your energy can be channeled into constructive transformative work, or it can be channeled into destructive revenge.

Progressed Mars Opposition Natal Ascendant

There is an increased likelihood of quarrels in this period.

Progressed Mars Opposition Natal Midheaven

There can be a failure caused by faulty planning, faulty motives, or involvement with the wrong people or goals.

Progressed Mars Opposition Natal North Node

There is an increased likelihood of quarrels or activity that isn't in the best interest of growth.

The most crucial periods in the progression cycle are the times when progressed Mars is within one degree of changing or having changed signs (or at 0 degrees of a sign), or when it's within one degree of changing or having changed house cusps (or on a house cusp), or within one degree of making an exact aspect to a planet, an angle, or the Moon's Node.

Sign and house changes are particularly important, as these are times when you become particularly aware of changes in your motivation and in the use of your energy. Depending on which progression system you use, the dates for sign changes will vary. This is why I suggest you experiment with different progression systems and see which one coincides most closely with the changing themes in your life. There will also be some discrepancy in the years your progressed Mars changes houses—not only because of the different progression systems, but because different house systems make for different house cusps. Only the angles remain unchanged. This is interesting, because traditionally, one of the prime significators of a major "turning point" in a person's life is a progression —particularly a conjunction to a natal angle.

As for whether or not Mars progressions can assist you in discovering the "best" house system for you, I have some doubts. I have a feeling that each and every house system represents things "as they are"—or maybe as you wish they were—at one level of your being. One house system may display your physical circumstances more graphically, one your mental development, one your emotional state, and so on. I've noticed that regardless of which system is used (assuming it's accurately calculated) progressed Mars crossing a house cusp seems to have led to a turning point on one level of my being—sometimes a change of situation, sometimes a change of interest or attitude. A few of my students who've checked out this theory have had the same sort of results. So it may be that while there's no best house system per se, there may be a house system you relate to better, according to whether your orientation is predominantly physical, mental, or emotional oriented, or whether you're self-motivated or environmentally motivated and so on.

For many years a debate has raged as to whether progressions symbolize inner changes or events. I think this is a question that can't be answered firmly. For an event to happen, there must be three factors present: first, you must be physically capable of participating in the event. This is shown by your natal potential. Second, you must be in a state of mind conducive to that sort of event. This is primarily where progressions come in, as these symbolize inner unfoldment and changes of awareness and motives. Finally, your environment must provide the circumstances or impetus for the event to take place. I feel this is where transits come in. The three factors must work together, and this is why transits, although they affect all people, may bring very different events to different people—or even very different events at different times to the same person. Though we may occupy the same environment, we're all born with different potential for coping with that environment and our potential unfolds in different ways and at different rates. For this reason, I feel that neither progressions nor transits can stand alone. To get the full picture, we need both.

Chapter 8

Transits of Mars

So far we've dealt with natal and progressed Mars. What about transiting Mars? What does it do? Transiting Mars energizes what it touches. It pushes you, motivates you, gives you a little kick in the bottom if you've been goofing off. It makes you more impulsive, more inclined to leap before you look into matters connected with what it touches. Mishandled, its energy can cause you to be excessively aggressive or demanding, which can lead to harsh words—or even threats, in extreme cases—being exchanged. Or it can lead to accidents. Energy is intensified with anything connected with the planet or house Mars touches. Overwork is common when Mars is "pushing" you; if you allow yourself to become exhausted, the danger of injury increases.

Transiting Mars in the Houses

The following guidelines can be used to interpret transiting Mars as it activates your natal chart.

Transiting Mars in the First House
You may become more ambitious than usual. If you're in too big of a hurry, there may be an accident. You're much more of a fighter than usual—especially if you feel your rights are being tampered with. You can become headstrong. There can be some sort of new beginning.

Transiting Mars in the Second House
Be practical in your spending or you'll be sorry later! On a more positive note, you can find or buy something that helps you accomplish a difficult task. You could become involved in work connected with machinery, psychology, or planting. Negatively, any natal tendencies toward greed are accentuated, and danger of theft is higher than usual while Mars is here.

Transiting Mars in the Third House

There's a restlessness, a tendency to want to take short trips just to be moving around rather than to accomplish anything. The pace of your everyday routine might speed up considerably. There may be a tendency to try to coerce others into believing what you believe. This is a good period for virtually any sort of mental work that requires a lot of energy. Debating the pros and cons of something may be a theme; the debate could get heated.

Transiting Mars in the Fourth House

This could be indicative of a fire in your home in extreme cases. Or it could merely indicate a great deal of activity where you live. Anger or an upset could lead to indigestion. There could be an injury in your home. Or there could be a family quarrel.

Transiting Mars in the Fifth House

You feel a strong need to get involved in some sort of physical activity. In searching for fun, you may go to extremes. You can become pregnant or impregnate someone. There can be a lack of sensitivity when dealing with children or lovers—watch that! There can be jealousy in the romantic area.

Transiting Mars in the Sixth House

Keep a firm grip on your temper or there will be problems. This can lead to a conflict with a superior. You may become involved in work that involves physical exertion or even danger. Or you might be involved in activities connected with engineering, appliance repair, or medicine.

Transiting Mars in the Seventh House

You may be involved in one or more enjoyable parties. Repressed annoyance or dissatisfaction can be brought out into the open and successfully dealt with. This can be a period when you put a great deal of energy into working with others, and there is a need for cooperation if you don't want problems. You could become involved in some sort of contest or competition.

Transiting Mars in the Eighth House

There may be trouble with joint finances, or someone close to you may have financial difficulties that somehow affect you. You may attend a function connected with the occult. You may be involved with insurance matters. There may be a battle with a big company. The sum of all your experiences in this period will lead to a big change even though the events at the time may seem fairly trivial.

Transiting Mars in the Ninth House

Be very careful when dealing with in-laws and/or foreigners; there could be friction. Under certain circumstances, this transit can indicate legal difficulties provided there are at least two other similar indicators present at the time. There is danger of fanatical behavior in connection with religious, educational, or minority group activities. You'd certainly be more assertive than usual when it comes to promoting your religious, philosophical, or ethnic beliefs. The most constructive thing to do with this transit is to put as much of your energy as possible into expanding your mind.

Transiting Mars in the Tenth House

A lot depends on what sort of aspects, if any, Mars is making. This transit arouses your ambition and your desire to achieve, but how much you actually accomplish depends on how active Mars is. Career matters are generally emphasized. Your mood is an enterprising one. Sometimes there is involvement with police, auditors, or psychologists.

Transiting Mars in the Eleventh House

You may receive invitations from casual friends or acquaintances. There can be energetic action involving friends or action in support of a humanitarian group. Or you may want to take on a group leadership role. This is an ideal time to formulate your goals for the future.

Transiting Mars in the Twelfth House

Don't let yourself be too impulsive or there will be trouble. You may be confined to hospital for surgery or laid-up as a result of flu or a minor accident. Or you may merely be irritable. It's best to work alone as much as possible during this period; you may instinctively realize this and become rather secretive and reclusive.

Transiting Mars Conjunct Sun

This increases impulsiveness, which in turn can cause your judgment to be faulty. You can feel that someone or something is holding your back; in reality this probably isn't the case—in fact, more than likely you've been goofing off and now want to make up for lost time! Tendencies toward authoritative behavior, anger, and rashness are increased.

Transiting Mars Conjunct Moon

This could be an emotional period. Your mood tends to be angry or excited. There can be activity connected with work, food, or gaining self-understanding. You may do something you didn't intend to do.

Transiting Mars Conjunct Mercury

Your mood tends to be more than usually argumentative, hasty, or extremist. There can be some sort of logical debate or a battle of wits.

Transiting Mars Conjunct Venus

The opposite sex finds you especially attractive. This is a favorable period for creative activities, particularly art, working with plants, or trying out new clothes or make-up or a new hairstyle. Negatively, you can lack consideration or moderation, which can lead to discord in the romantic area.

Transiting Mars Conjunct Mars
Watch your tongue! Energy is increased, but can be wasted if not consciously channeled. Ruthlessness can lead to conflicts.

Transiting Mars Conjunct Jupiter
Think twice before you act; the tendency to get carried away is high. Enthusiasm, enjoyment of life, and enterprising tendencies increase. This is a favorable period in which to look for work.

Transiting Mars Conjunct Saturn
This could be a frustrating period. There is a tendency to suppress your anger; perhaps you have no choice. There can also be a tendency to do (or have done to you) bitchy little things on purpose rather than openly complain or disagree. Any natal tendencies toward severity or harshness are increased.

Transiting Mars Conjunct Uranus
This could cause trouble in a partnership or other relationships. Or it could merely have you taking the initiative in connection with joint finances or dealings with a big business. Both physical energy and irritability tend to increase. Your mood and actions are contradictory.

Transiting Mars Conjunct Neptune
This is apt to be a confusing period. Secretive tendencies increase, and you could be involved with spiritual healing, past life recall, or some idealistic philosophy. You may want something but not want to do anything to get it. Energy could be misused.

Transiting Mars Conjunct Pluto
Don't let yourself get too aggressive or there will be trouble. Assertive, ruthless, and brutal tendencies are strengthened. It would be wise to avoid potentially physically dangerous situations.

Transiting Mars Conjunct Ascendant
The theme of this influence is to assert yourself in a way that affects others because you have a chance to show those in your environment what you can do. Impulsive tendencies and aggressiveness increase. There also can be conflict with someone in your environment.

Transiting Mars Conjunct Midheaven
There's a tendency to act impulsively; watch that! You could take dynamic action in connection with politics, money, psychology, your career, or your status. Your mood is excitable, and there's a tendency to act without thinking.

Transiting Mars Conjunct North Node

You may be overly optimistic during this time. Also, you may compete for recognition or social acceptance, such as looking for social approval of your physical condition that can include dieting, buying flattering clothes, participating in sports, et cetera. Cooperation may be hard to get, and you may have to ''bribe'' someone to go along with what you want. A relationship can be disrupted by circumstances more or less beyond your control.

Transiting Mars Semi-sextile Sun

This transit can make you slightly more egotistical or impulsive than usual or bring a minor irritation connected with the involved house or houses.

Transiting Mars Semi-sextile Moon

You may be annoyed with the weather, and slightly more rash or rebellious than usual.

Transiting Mars Semi-sextile Mercury

There can be a minor irritation connected with work or studies, and someone could harp about a trivial matter. You'll be slightly more restless than usual.

Transiting Mars Semi-sextile Venus

Someone can meddle in your work or private life. You can become overly excited or slightly tactless.

Transiting Mars Semi-sextile Mars

For some reason you can't take direct action to achieve what you want; you have to deal with an intermediary or act in a roundabout way. This slightly lessens inner calm, and can also be a factor in violent acts, although it's too weak to trigger violence on its own.

Transiting Mars Semi-sextile Jupiter

There can be snags or red tape of a minor nature if handling legal matters. You want to assert yourself but can't because the time isn't right. There can be minor but annoying events connected with meetings or negotiations.

Transiting Mars Semi-sextile Saturn

Determination doesn't seem to get you far. You may have to submit to someone or something you don't want to. Stubbornness increases slightly, and any tendencies to procrastinate may also increase.

Transiting Mars Semi-sextile Uranus

Willpower is erratic. There is inner turmoil or tension; you want to release this but for some reason are unable to do this.

Transiting Mars Semi-sextile Neptune
Occult activities hit minor snags. Moodiness and/or feelings of inferiority may be slightly stronger than usual.

Transiting Mars Semi-sextile Pluto
There can be half-hearted action connected with the occult. Violent tendencies increase slightly as does a tendency to be accident-prone, but this aspect in itself isn't strong enough to trigger violence or accidents.

Transiting Mars Semi-sextile Ascendant
There can be problems with sharp instruments, or a minor quarrel. If you're a physical person by nature, this influence slightly increases the likelihood for a fight.

Transiting Mars Semi-sextile Midheaven
There can be minor conflict with an authority figure or at work. Aimlessness increases slightly.

Transiting Mars Semi-sextile North Node
Be practical in whatever you do or there will be snags. Actions tend to be out-of-sync; timing is slightly off. Quarrelsome tendencies increase slightly.

Transiting Mars Sextile Sun
People in high places tend to be more than usually receptive to you. You're more energetic than usual, and you assert yourself in connection with your ideas. Resolve is strengthened. Creative abilities are enhanced.

Transiting Mars Sextile Moon
You may become involved with a member of the opposite sex. Or you may be participating in or attending a sports function with one or more family members. Or you may become aware of a goal for the first time. Your mood is changeable, emotional, impulsive.

Transiting Mars Sextile Mercury
You can think faster than usual, and tend to assert your ideas in a positive manner. You could turn an idea into reality or prove a point. There will be more than the usual amount of mental activity and communication.

Transiting Mars Sextile Venus
If you're fun-loving, outgoing, and/or highly emotional, your desire for fun could cause you to ignore your work, or your emotions could somehow get the better of you. Otherwise, you'll take more initiative than usual in the social area or maybe take a short trip for social reasons. Your need for love is stronger than usual, and sensual tendencies tend to increase.

Transiting Mars Sextile Mars

This can be an excellent period if you're a relaxed person by nature; if you're not, it can make you a trifle hyper. This aspect increases your energy and can help you accomplish your goals. Effective action can be taken with, or in connection with friends. Determination is increased. In general, there'll be more activity than usual.

Transiting Mars Sextile Jupiter

This is an excellent time to form a business partnership. It's also favorable for most other work-related matters. You can become enthusiastic about a worthwhile educational program and take action to become part of it. You can form a link with some helpful organization. You become determined to act in some area.

Transiting Mars Sextile Saturn

Now you begin to get results from your efforts. Efficiency, willpower, and powers of concentration are increased. Work-related matters should go well.

Transiting Mars Sextile Uranus

A superior may help you get a promotion, raise, or something else you've been trying to get for months. Resourcefulness, creativity, and your need for freedom increase. There can be success in terms of achieving a goal.

Transiting Mars Sextile Neptune

Don't be deceived by appearances; things may not be what they seem to be. You can receive intuitive guidance in the work area. Actions tend to be intelligently directed, but if idealistic tendencies are strong in your natal chart, there may be a bit more optimism than is actually warranted. Energy tends to be more intellectual than physical. Spiritual tendencies are increased; religion or the occult attracts.

Transiting Mars Sextile Pluto

You're more energetic than usual. You could become interested in investigating some scientific, psychological, or occult subject. You could take some sort of constructive action connected with joint finances, work, or dealings with a big company. You can achieve a small success. Your enthusiam for what you're doing is high; if you're an enthusiastic person to begin with you can even become a trifle fanatical.

Transiting Mars Sextile Ascendant

You could be busy with friends, energy increases, and you actively cooperate in an important relationship. If you're 'bossy' by nature, you might try to force others to do what you want. You want to lead in any case.

Transiting Mars Sextile Midheaven

This is a period during which you normally have a considerable amount of self-confidence and healthy ego needs. Your ability to assert yourself is especially good. You'll probably want to act on your own rather than with others.

Transiting Mars Sextile North Node

You may have to make an important decision during this period. You can be involved in cultural activities or sports. Cooperation and companionship are favored.

Transiting Mars Square Sun

Watch your tongue or there will be trouble! Aggressiveness, anger, obstinacy, and impulsiveness are all stimulated.

Transiting Mars Square Moon

You could be impulsive, irritable, or quarrelsome. Family problems may interfere with work. You may do something you didn't intend to do.

Transiting Mars Square Mercury

If possible, postpone making decisions until this has passed, especially if they pertain to finances or any sort of irrevocable change. There can be back-stabbing or a tendency to bite the hand that feeds you. Your mood can be caustic; you can be looking for a fight. Nervousness tends to increase.

Transiting Mars Square Venus

Carefully watch your finances as there's a strong tendency to overspend in this period. There can be problems in the romantic, marital, and/or social areas, including a one-night stand. In general, you tend to be dissatisfied with whatever happens during this period—even if it's positive.

Transiting Mars Square Mars

Don't try to force or rush things; there's liable to be an argument if you do. Impulsiveness, argumentativeness, and any tendencies toward destructive behavior are increased. There can be conflict over who's in charge or conflict with an authority figure.

Transiting Mars Square Jupiter

There could be an unexpected expense of a financially draining nature. There can be a tendency to do a guru trip if you teach religion, metaphysics, or philosophy. Fanatical behavior can lead to conflict, and there can be difficulty with a superior.

Transiting Mars Square Saturn

Your mood tends to be argumentative; you're hard to get along with. There could be work-related hassles, and in extreme cases you might even be fired. You could have a particularly heavy work load. You want to overcome an obstacle and aren't too discriminating about how you do it. If a relationship (romantic or otherwise) has been going badly for a while, this may be the time it ends.

Transiting Mars Square Uranus

This tends to be a frustrating period during which unexpected events tend to throw you off balance. Egotistical tendencies increase, there can be disagreements, and your nerves are tested. Maybe you'll have to deal with someone or something you detest. There is an increased tendency to be accident-prone.

Transiting Mars Square Neptune

Be extra careful if you're around water; boating and swimming mishaps as well as scalds can occur. You can be deceived in the work area or defrauded or "ripped off" by an insurance company, con man, or thief. There can be a disappointment. In extreme cases you can be harmed physically or psychologically.

Transiting Mars Square Pluto

Don't let the aggressiveness this influence stimulates provoke an argument. You might try to force someone or something to be what it isn't, but caution is necessary if you don't want problems. An accident-prone tendency and ruthlessness are increased.

Transiting Mars Square Ascendant

This generally brings a stormy period in the relationship area. Lack of consideration can lead to conflict. Your mood is aggressive. You may even be "asking" for a fight of either a physical or verbal nature.

Transiting Mars Square Midheaven

In this period you're apt to feel quite aggressive. There can be a conflict in the work area, and an ambition can be frustrated. Your mood is excitable; you tend to act without thinking.

Transiting Mars Square North Node

You may find it harder than usual to borrow money or make ends meet. Your actions clash with the culture or attitudes of those around you. Poor timing affects work and other areas. Cooperation is hard to come by. A relationship can be disrupted; plans can be canceled or postponed.

Transiting Mars Trine Sun

This influence can indicate improved health and/or energy level. It increases courage, dynamism, leadership ability, and ambition, and also sharpens your instincts.

Transiting Mars Trine Moon

Unlike most of the other Mars-Moon transits, your emotions tend to be fairly stable. Energy increases, and you're less timid than usual, and more inclined to act constructively on your feelings. Enterprising tendencies are enhanced.

Transiting Mars trine Mercury

This is an excellent time to answer any correspondence or make phone calls you've been putting off. Decisiveness, mental abilities, and enterprising tendencies are enhanced. If you want to settle something quickly, this is the time to do it!

Transiting Mars Trine Venus

This should be a good period for creative people, particularly artists, chefs, florists, and hair stylists. It's also a favorable period for becoming pregnant. And it's good for actors, psychologists, and dancers, too. Whatever creative ability you have is enhanced. Wanting something badly and trying to get it could also be a theme during this time.

Transiting Mars Trine Mars

Energy level and resistance to illness are higher than usual. You could enjoy sports or some other type of physical activity now, and this is a favorable time for work-related travel. In general, you enjoy your work more than usual. Assertiveness is increased.

Transiting Mars Trine Jupiter

This can be a lucky period; timing is very good. It's an excellent period for settling legal matters, and also a good time for work-related travel. Enthusiasm and creativity increase.

Transiting Mars Trine Saturn

Work should go well. Energy level is good, and a health problem may improve. Ambition and perseverance are intensified, and as a result you tire less easily than usual.

Transiting Mars Trine Uranus

Your intuition could be more reliable than usual, and you're constructively adventurous. No matter what happens, you feel you can handle it, and you're able to make decisions quickly.

Transiting Mars Trine Neptune

You may study philosophy, alchemy, or teachings from a metaphysical group (Rosicrucians, Masons, Brotherhood of Light, et cetera). Intuition is stimulated as you're more receptive than usual. You can receive intuitive guidance in work or some other area of your life. Energy tends to be more intellectual than physical.

Transiting Mars Trine Pluto
Energy level is good, and there could be an improvement in the health area. You could assume a leadership role at work or in some other area. You tend to use joint finances and credit effectively. There could be a small success. You're enthusiastic—maybe even fanatical—about what you're doing.

Transiting Mars Trine Ascendant
This is normally a period when you feel self-assured. You'll want to enlarge the scope of your activities to some extent. Action in the relationship area tends to emphasize the intellectual a bit more than the physical. You could be working with others or working alone to bring about a success.

Transiting Mars Trine Midheaven
You undertake whatever you have to do with a great deal of enthusiasm. Your energy level is higher than usual. Family activities tend to be more intellectual than physical. Your decision-making abilities are especially good, and you could become aware of a goal for the first time.

Transiting Mars Trine North Node
If you're inclined to rush, this will make you really frenetic; otherwise it merely speeds you up. You have greater than usual appreciation of intellectual subjects, and may be involved in cultural activities. Dealings with others are likely. You may want to have children.

Transiting Mars Quincunx Sun
Determination either wavers or turns to stubbornness. Quarrelsome tendencies increase. There's a higher than usual vulnerability to illness.

Transiting Mars Quincunx Moon
On the singles scene, there can be sexual attraction without any other sort of rapport to back it up. Your mood, regardless of your romantic status, tends to be excited or rash.

Transiting Mars Quincunx Mercury
Motivation is either erratic or high, but you're out of touch with something and don't know it. Tendencies toward hastiness and extremism are increased.

Transiting Mars Quincunx Venus
It's possible you'll receive a gift that doesn't fit or something you don't want; it might have strings attached to it. You might feel blah or have a minor health problem. There could be a disagreement with a member of the opposite sex.

Transiting Mars Quincunx Mars

Progress is interfered with because a task requires more physical exertion than you feel capable of. Energy can be wasted. There's a tendency to overreact to people and things.

Transiting Mars Quincunx Jupiter

Your intentions are constructive but your actions are inappropriate. There's a tendency to act without thinking. You want to assert yourself so badly that you blurt things out regardless of whether others are in a receptive mood.

Transiting Mars Quincunx Saturn

You want to accomplish something practical, but your discipline is erratic. A tie may be severed, or you could be separated from someone you care about. There can be grief about circumstances beyond your control, or merely a tendency to cry over spilled milk.

Transiting Mars Quincunx Uranus

You can overdo exercise or physical exertion, resulting in strains or sprains. Machines can also be hazardous to your health. Plans can be spoiled.

Transiting Mars Quincunx Neptune

This is not a good period for research as you tend to get sidetracked by interesting irrelevancies. You can give someone the benefit of the doubt to your own detriment, and someone could trick you or talk negatively about you behind your back.

Transiting Mars Quincunx Pluto

You receive money, and the next day you get an unexpected bill. A refund or check you're expecting may turn out to be less than you thought. Tendencies toward brutality are increased, and this aspect can contribute to involvement in violence.

Transiting Mars Quincunx Ascendant

There could be an illness, accident, quarrel, or conflict.

Transiting Mars Quincunx Midheaven

Physical activity can be overdone, leading to strain. You can overshoot your mark in some area and have to backtrack. There can be a conflict in the career area.

Transiting Mars Quincunx North Node

Be extra careful around old buildings, sharp instruments, and fire during this period. And don't go overboard with sports or physical activities that are unfamiliar to you or infrequent parts of your lifestyle. Quarrelsome tendencies are increased, as is the tendency to make mountains out of molehills.

Transiting Mars Opposition Sun
It seems like half the people you know are opposing your plans. There can be power plays in the work area. Or you need to cooperate with an authority figure even though you don't agree with this person. Someone can make outrageous demands on you. Occasionally this triggers an illness or an accident.

Transiting Mars Opposition Moon
Be as tactful as possible in oral and written communication or there will be problems. If driving, observe the speed limit; this transit has been known to trigger tickets and, occasionally, minor fender-benders. A compromise or sacrifice can be necessary in connection with a family matter. Rebellious and quarrelsome tendencies tend to increase.

Transiting Mars Opposition Mercury
This aspect can trigger tension. Your temper can get the better of you if you don't consciously control it. Arguments are common. Negotiations involving work or professionals tend to be difficult. Restlessness and nagging are frequent manifestations.

Transiting Mars Opposition Venus
Your feelings may be supersensitive, so much so that you're hurt by some trivial little slight or unintentional remark. There can be jealousy, excessive aggression, or excessive emphasis on sex. Love life or marriage could be discordant.

Transiting Mars Opposition Mars
Curb your impatience if you don't want problems. Egotism is common; so is feeling a need to be in charge. Battles can occur. There's a lack of inner calm. In extreme cases, there can be violence.

Transiting Mars Opposition Jupiter
Closely watch your budget as extravagance is common in this period. If self-employed, there's a tendency to over-expand; if working for others, their expansion can cause you headaches. Your tendency is to push for a decision, and you may be willing to settle for less than you deserve just as long as you know where you stand. You could meet with or negotiate with someone who doesn't agree with you.

Transiting Mars Opposition Saturn
You may be trying to go in two different directions—simultaneously. Or you could be so fed up you just want to get away from everything. There can be excessive physical strain. You could break a bone, become ill, or have trouble with your teeth. Any tendencies toward severity would be increased.

Transiting Mars Opposition Uranus
You may be depressed. Your mood is rash, irritable, contradictory. There can be conflict with a friend.

Transiting Mars Opposition Neptune
Watch your tongue; your emotions may cause you to say things you'll later regret. You can contribute to your own self-undoing by being sneaky or plotting to make someone regret their actions. You can be defrauded, taken advantage of, or ripped off. Some physical abnormality may be discovered. This is not a good time to indulge in drugs; there can be bad consequences.

Transiting Mars Opposition Pluto
Watch your temper! You're more aggressive than usual, and there can be an increased tendency for conflict as well as accidents. There can also be a subtle sort of harmful activity going on.

Transiting Mars Opposition Ascendant
This is an extremely tacky transit in my experience. It seems that even partnerships and close relationships that run quite well normally can be problematical in this period. Dealings with dynamic, aggressive people are likely. There can be a conflict with someone in your environment.

Transiting Mars Opposition Midheaven
Unless this coincides with transiting Mars in a conjunction with natal Mars, this influence is very similar to becoming invisible; nobody notices you. You tend to anger more easily than usual. There can be an accident in your home. Subconscious emotions are aroused, and these may lead to upsets. Your mood is excitable.

Transiting Mars Opposition North Node
You may be overly optimistic, and take action to support a conservative status-quo-loving person or cause. You can be unpopular at work or in some other area—probably because of a refusal to change something. Cooperation is hard to achieve without a fight, compromise, or sacrifice. A relationship can be disrupted.

Remember that transiting Mars can also station on a planet, turn retrograde, and then turn direct again. When you see this happening, know that there's something important going on, that there may be a temporary disruption or break in the action, and that you may have to do things more than once to bring about the desired results. This may be frustrating, but generally it goes with the territory of life—and if you see it coming, you can keep hassles to a minimum.

Chapter 9

Your Mars Return Chart

Have you ever noticed that there seems to be about a two-year pattern in your own life and other people's? You may be heavily involved in doing something for a couple of years and put all your energy into it, then suddenly you're just not there any more. Your motivation flags. Something else catches your fancy. Something better comes along. And off you go. Have you ever wondered why? Well, in many cases, the answer lies with transiting Mars and can be found in the Mars return chart. The Mars return chart is a special kind of transit chart made for the date, time, and place, when transiting Mars returns to it's natal position, which it does roughly every twenty-two months. This chart can be calculated using almost any standard astrological software program. It is a useful tool in terms of showing you what you're liable to be doing during the next couple of years, and can be particularly useful in terms of seeing job changes and other major lifestyle changes.

Here's how you work with it:

- First of all, remember that this is a transit chart. Transit means transitory, so this chart has a limited shelf life, but during that period it can offer important clues about where your energy is going, what will be turning you on, and what you're motivated to start—or in some cases, cut out of your life.
- Set up the chart with an aspectarian, just as you would for a natal chart, solar return chart, or any other chart and we'll walk through this together.
- Note that the Mars position is identical to that of your natal Mars. The Mars return always reinforces your natal Mars potentials, so the natal position is always activated.
- Now look at the house position of Mars in the return chart. This is going to be an area of

emphasis over the next twenty-two months or so, and you can expect to be "doing" something there, as follows:

First House

You're getting a start on something and are probably hoping to be first in line. The "something" will vary, but will probably be connected to the rising sign in your return chart. You'll probably have plenty of energy to do what you want to do; in fact, you may very well have more energy than you know what to do with! You could have lots of plans and ideas, and you could very easily give every single one of them a try—but remember, Mars often has a tendency to start more than it finishes. While you may very well feel like Superman or Superwoman and think you can do it all, in reality you're only one person and no matter how much energy you have, there are still only twenty-four hours in a day and some of them are going to be taken up with boring stuff like eating, sleeping, and doing the laundry. Suffice it to say, an attempt to prioritize your projects wouldn't hurt!

Anger is the bugaboo to watch out for this year, especially if you're one of those sweet souls who tends to hold it all in or turn it all inward most of the time. You can expect to be assertive this year and you can expect yourself to be disinclined to take anything remotely resembling abuse from anybody. And if somebody's been dishing out the bully barbs for a while and you've been just taking it, they may be in for a surprise as this is very likely to be a time when you'll fight back or cut the tie and walk away. In some instances, this could be a very good thing, but keep in mind that Mars here can incline you to be reckless and occasionally self-destructive. There are times when telling someone to take the job, the marriage, or whatever other carrot they are dangling and shove it is in your best interests and indeed absolutely necessary; there are other times when you are simply cutting off your nose to spite your face. The aspects to Mars in the return chart—which, by the way are also the transiting aspects to your natal Mars (not to mention the rest of your chart)— will give you a good idea of which scenario is which.

Second House

If you're working, expect to be hustling. Things will be busy, and yes, money will be coming in at a good clip, assuming you keep up with everything. In general, this suggests a higher income, but I've occasionally seen exceptions. One of my clients, a self-employed consultant, got up the day before her Mars return and realized that she had been working, on average, sixty hours a week for the past two years, had not had a vacation in five years, and really needed a break. She decided to take fewer clients—which did mean a pay cut, but also gave her more time to do other things she enjoyed. Another client had Mars conjunct Saturn in his Mars return chart. His company was hit hard by the recession and he went from a forty-hour work week with occasional overtime, to a twenty-hour work week and occasional calls telling him there was no work and not to bother coming in. Remember, Mars can at times cut things. It all depends on your aspects and your current situation.

If you're not working, this could instead mean that your value system is changing. This might mean that the emphasis on material things or keeping up with the Joneses may change one way or the other. In particular, if there's something you've really, really, wanted that everyone says is not worth it, this could be your time to go for it or give up on the idea.

Third House

Mars is not elegant, gentle, or sweet. Mars is assertive and calls a spade a spade. And that's what you'll be doing too. Your mind will be sharp; your tongue will be likewise. It's not too likely anybody's going to pull the wool over your eyes this year—and if they try, they may live to regret it! You can expect to have lots of experiences that lead to new insights. You can also expect those new insights to lead to action and probably action will take you in a new direction or otherwise lead to change.

What should you watch out for? Well first of all, you might want to look before you leap. The tendency to jump to conclusions is increased during this period. Second, be aware you might tend to lose your temper a lot more quickly than usual. If you're going to blow your top, make sure it's over something that's worth the effort. Last but not least, if you drive, watch the speed limit. For those who insist on placing a lead foot on the gas pedal or not coming to a full stop at the stop sign, tickets or even fender-benders could be waiting in the wings.

Fourth House

You're apt to be expending a great deal of energy working around your home. You could be redecorating or renovating, but you could also simply be cleaning out the garage, the basement, or the junk room and getting rid of stuff. And once in a while this actually means you're moving out and moving on. In general, the work you're doing is demanding and is often of the sort where one thing leads to another. For example, you decide to paint the bathroom and when you're done you decide that the flooring looks tacky so you decide to replace that too.

Mars is anger. In the fourth house, it is often anger that's been bubbling for a while, so airing old resentments and rehashing past hurts is not uncommon. This anger is invariably directed at parents or your significant other, though once in a while it's a roommate. Sometimes airing these old wounds can be very healing and can allow closure. Sometimes, however, the door is closed literally. "Leaving home" scenarios are not uncommon, and while sometimes they involve leaving home because you're going off to school or getting a job in another state or province, I have also seen a fair number of leave-takings involving slammed doors during this one. Do what you need to do, but don't cut off your nose to spite your face.

Fifth House

Time to have fun, fun, fun! Self-restraint? Not in fashion at the moment. Going out on a limb? Yeah, that's more like it. New hobbies, new passions, and new trains of thought are all very common—and chances are that you'll dive into them wholeheartedly. This isn't necessarily a bad thing—trying new things can increase your confidence, especially when you succeed or at least discover that you like what you've tried. You will be direct about what you want, and you'll probably also insist on the right to be free to do what you want and pursue what you want. Self-assertion is generally the primary mode of self-expression in this period.

The fifth house is often connected with romance, so not surprisingly romantic issues can arise. Friends can become lovers, lovers can become ex-lovers, and pretty much anything in between can happen. Since the emphasis is on fun, this is not necessarily a time of balance in terms of your love life. And yes, there can be arguments as a result. Last but not least, if you have children, expect them to keep you particularly busy.

Sixth House

This works best if you're self-employed or in a position where you work largely on your own with a minimum of supervision. You're definitely motivated to get the work done, but you want to do it your own way at your own pace—and you're probably not going to suffer fools gladly. Oh, and did I mention that probably anyone who doesn't want to do it your way is apt to be seen as at least a bit of a fool? Uh-huh. You can be angered more easily than usual if others slow you down, whether it's because they aren't returning phone calls or because they're hogging the photocopier when you need it. Whatever.

A little caution is needed on the health front with this one for a number of reasons. First of all, Mars is connected with accidents. If your work—occupationally or otherwise—involves sharp implements, heavy equipment, or climbing a ladder to get stuff that's stored on high shelves, a little extra care is in order, especially if Mars is under stress. Second, while you're motivated to get things done and your energy level is probably good, burning the midnight oil is a distinct possibility. Burn enough of that oil and you could be heading for a burnout—and lack of sleep is another thing that can contribute to accidents. Last but not least, if you have bad habits that impact your health, this may be the time when you're given the ultimatum to cut it out if you want to be around for many more Mars returns. Drinking and smoking are the most likely vices to need cutting out, but there are other things that may need to be looked at—cholesterol, use of energy or diet supplements, or pain killers come to mind. Once in a while, some kind of corrective surgery needs to be done. At the very least, a check-up is probably in order if you haven't had one for a while—even if you feel fine.

Seventh House

You're meeting and dealing with people who display Mars behavior—assertiveness, enthusiasm, and high energy for starters. Hopefully they will be people who display the positive side of Mars—and hopefully they'll inspire you, motivate you, and have your best interests at heart. They may offer you leadership, social opportunities, job opportunities, or—dare I say it—even sex. They may, on some level, be life skill coaches who can teach you something valuable about yourself or about life in general.

As usual, though, there's another option that may not be as palatable. They may just make you angry. Unfortunately, this isn't necessarily simply a case of some jerk who cuts you off in traffic or pushes in front of you in line. It could actually be someone that you deal with on a daily basis—a spouse, a business partner, or someone who, up until now, has been a close friend. Now they start to get on your nerves. And maybe you're getting on their nerves too. If someone is making you mad every time you see them, maybe it's time to say something. With Mars, assertiveness is always the antidote to anger. And if a straightforward discussion does nothing to dispel the angry feelings, perhaps it's time to move on.

Eighth House

The eighth house is connected with some of life's relatively non-negotiable aspects—death, sex, and taxes, for example. It's also connected with other people's values. I have found that when Mars is in the eighth house in a return chart, one of these inevitables meets with resistance. Take sex, for an example. Sometimes this is a period when you encounter some very strong chemistry but for one reason or another are ambivalent about exploring it. Maybe you're already married or otherwise involved. Or maybe this person is, at first glance, not your type—by which I mean he or she doesn't fit the superficial stereotype of what you think you're looking for. What does it mean? It means you have some soul-searching to do. Sometimes instead there is a need to confront your own or other people's mortality; it is regrettably not uncommon to hear of someone's passing or of someone being diagnosed with a terminal illness in this period. And sometimes this is simply a reminder that you need to somehow pull your weight or pay your share. So if you haven't been filing your taxes, expect the tax man to find you. And if you haven't been contributing your fair share at work or at home, expect to be put on notice to clean up your act!

The eighth house is where we bury all the stuff that is too ugly or too painful to deal with. Those of us who try to follow a spiritual path are particularly prone to forgiving in the name of spirituality. The problem is that while we may forgive, we generally do not forget, and while we may wrap our anger in layers of love and light and put it in a safe hiding place, the pain of past experiences doesn't always go away. You may be upset with others for things that happened in the past—and these could be very major transgressions or fairly minor ones. If the relationship is already strained, this can be a time when you rehash every single instance of misbehavior from day one in an attempt to seek redress. And indeed if you want these transgressions to stop, it may definitely be time to do that. Or in some instances it may be time to seek counseling of some sort in order to let go of the pain and move on.

Ninth House

Generally this is the time when you make a commitment to some sort of ninth house endeavor. You may enter a college or university program or some sort of program of specialized studies. You may decide that this is the year you're going to take that trip overseas. Or you may decide that it's time to make a commitment to your spirituality. One of my clients spent this Mars return period on a kibbutz. Another became a reverend in a spiritualist church. Most people who experience this return encounter exciting and inspirational people who lead them to new—and sometimes greener—avenues of fulfillment.

Once upon a time the ninth house was connected with foreigners. It still is, but in these times of global travel and emigration, it's sometimes hard to tell who is and who isn't a foreigner. If you were born in England but have lived in the US for the past twenty years, are Americans still foreigners? If you routinely travel between here and Australia, is Australia still foreign? My take on the ninth house and the "foreign" issue is that foreign is, at any given time, what is culturally or territorially unfamiliar. And that brings us to the potential trouble spot with this return. If you are very fixed or for some reason have not learned to cope with common cultural or religious differences, or if your normal way of coping with these differences is to shoot them down or cut them out of your life, this could be a tricky time for you. Expect to have your prejudices and biases confronted—and expect to be learning something about religions or cultures other than the ones you were brought up with—whether you want to be enlightened or not! And if you were taught to never deal with certain segments of society because of race, color, or creed, expect to have those teachings challenged and in all probability proven wrong!

Tenth House
You can accomplish a lot this year! You can forge ahead in your career and otherwise make a mark on the world. You can gain status as a leader, a mover, or a shaker. You become more assertive. You also go after what you want. You may have to work harder than usual, but if you're willing to put forth the effort, chances are good that you'll end up this cycle ahead of where you started it. And yes, generally the emphasis is on career ambitions, though sometimes they are not your career ambitions but instead are the ambitions of a parent or spouse. Either way, these are apt to be ambitions that impact your reputation and/or your role in your environment. And yes, they can definitely involve a job change as well as more exposure in your little corner of the world.

Negatively, there could be some interference with or disagreement with your goals. What was said about the sixth house in terms of not suffering fools gladly and having no patience with people who hold you back or get in your way is also a strong theme with this return position. If you are self-employed, work for yourself, or otherwise have the final say in decisions impacting your status and reputation, you should have fewer problems—though if there are stress aspects to Mars, this period may not be entirely smooth sailing. If, on the other hand, you're in a situation where others call most of the shots, this can be a frustrating time—and if you were unhappy going into it there could very well be a "take this job—or title or whatever—and shove it" scenario. And again, for better or worse, what you do will be noticed, so make sure your actions truly reflect what you feel is for the best.

Eleventh House

This is a period when you want to keep your options open. You won't let anyone cramp your style if you can help it. You don't want your goals for the future restricted in any way. You'll assert yourself and protect your right to proceed as you see fit. Friends may shake their heads, but they can't stop you. You're going to do what you're going to do anyway.

Sometimes we set goals for ourselves that are beyond our ability to achieve singlehandedly. Sometimes we need help with what we set out to do. In this case, friends, professional associations, or other sorts of networks may come to our aid. While this is good, after a time it may turn out that this help had strings attached. Other times, our ties to others cause us to abandon goals out of deference to other people's wishes. If this is a time when you truly need help to start or accomplish something, rest assured that the help will be there. If, on the other hand, those strings or ties are now tripping you up and keeping you from the next phase of your journey, this could very well be when the ties are cut.

Twelfth House

Where you're headed is up to you. If you don't know where you're going, the best direction to go is an inward one. That's where the answers are. Keeping a diary, meditating, activities such as yoga or t'ai chi or even counseling may help clarify your path, but be aware that the path is probably a solitary one and this year is most likely the road less traveled. Also be aware that it is probably a path that's going to require attention to body, mind, and spirit.

Temper is often placed in the closet for the duration of this cycle. Even when it isn't, it normally doesn't do much good. Sure you may get mad. You may even lose your temper. But even while you're berating others, there tends to be a little voice of wisdom telling you that mostly you're mad at yourself. Why? Probably for not being who you really are on some level. Or possibly for forgetting who you are. There tends to be a process of reflection and often reclaiming or recovery in this period; it tends to be slow, but it's also absolutely necessary. During this process, you make valuable discoveries about your own adequacy, and while you may find it lacking in some respects, there's also a very good chance that by the end of the cycle you will have discovered strengths and skills you didn't know you had. So while this may not be as exciting as some of the other returns, in retrospect it often proves to be every bit as rewarding if not more so.

To recap, at the time of the return, Mars is always in the same sign and degree as it was when you were born. The aspects at the time of return are also the aspects to your natal Mars and to the rest of your natal chart. And these transits denote a cycle that lasts until the next Mars return and is an underlying theme for the twenty-two months or so. Except, of course, if Mars happens to be retrograde at the time of your return. Which does happen sometimes. In this case, there can be a number of additional factors to take into account. If Mars returns and then later retrogrades and returns to its natal position, sometimes it's an indication of a false start. This might mean that you start something and then change your mind about it and decide to do something else. Or it might mean that you start something only to discover that something that was left undone—or hasn't been done at all—is tripping you up and you have to go back and deal with whatever it is before you can move forward. Once you do move forward, it can be along the same lines as the first return or it can be in a somewhat different direction. Either way, once Mars goes direct, you begin to move forward and do what needs to be done in a straightforward way.

Here's wishing you happy returns!

Chapter 10

Mars and Your Inner Motivation

So far we've looked at Mars and what it symbolizes in your outer life—your self-expression, your relationship potentials, your career potentials, and so on. Now we're going to take a look at how Mars affects your inner life—your motivation for being here, your psycho-spiritual stamina, and, perhaps, the impact of your past lives on your present circumstances.

Exoterically, Mars is the ruler of Aries, sign of new beginnings. Esoterically, Mars is the ruler of Scorpio, sign of death, rebirth, and transformation. What does this mean? I think it means that Mars can be used to find a thread of continuity between this life and a previous one; not, perhaps, the most recent previous one, but definitely one that has a bearing on our present life nonetheless. To me, any house with Scorpio on the cusp shows unfinished business or a need for transformation of attitudes or skills—karma, in other words. Likewise, the house cusp where Scorpio is shows a place where you must pick up where you left off in terms of the life-circumstances symbolized. Any house cusp with Aries on it, on the other hand, suggests that this is where you'll be breaking new ground or involving yourself with new people or experiences. Should either of these signs be both intercepted and devoid of planets, you'll know that at one level of your being, at least, there are no totally alien life-experiences this time around (in the case of Aries) or no unfinished business of urgency (in the case of Scorpio). Note, however, that I said "at one level of your being." I do feel that each house system has relevance on one level of our being, and different house systems give different intercepts. Further, Equal House has no intercepts. So on one level of your being—at least if you buy my theory—we all have unfinished business as well as new experiences to look forward to.

In the case of the various "unequal" house systems, just as it's possible to have an intercepted sign, it's also possible to have a repeating sign; that is, a sign that shows up on two house cusps. Should this sign be Aries, then at one level of your being there's a strong need to do it yourself, go it alone. There may be hard work at this level, but the only one who can impede you is you. If, on the other hand, Scorpio repeats, then at one level of your being other people will figure very heavily into your life because Scorpio has to do with other people and their resources, what they owe you and what you owe them. At this level of experience, you may at times get valuable assistance from others, but you may also have more than your share of people-problems of some sort.

That brings us back to Mars as the thread, the key, to what we're doing here. To me, the sign holding Mars says a lot about your motivation for being here. And, as the exoteric ruler of Aries and the esoteric ruler of Scorpio, it may hold clues to what we've left unfinished (Scorpio) as well as what we've yet to experience (Aries), things that in short are somehow attracting us to good old planet Earth once again.

Mars and Past Lives

The following descriptions are meant to serve as psychic "memory-joggers," and are not meant to be firm and irrevocable laws. You may, as you read, be disappointed in your particular goal or feel it trivial. Don't be. For one thing, it's impossible to go into all the ramifications of your particular life in one short description because your chart is unique to you. Books, though good, can only give general pointers or guidelines. For another thing, in terms of evolution, no goal or action is of itself any better or worse than any other; it's merely something you feel a need to do. Your path, assuming you're following it and you're not over at the "evolutionary rest area" having a picnic or out of gas in the middle of the road and holding up traffic, is ideal for you. It may be a lousy path for your mate or best friend, but that's okay. That person's path is apt to be an equally poor choice for you. In astrology—especially esoteric astrology—the goal is not to judge but rather to understand. With that said, here are some guidelines.

Mars in Aries

You've developed a great deal of assertiveness—maybe even aggressiveness—in past lives. You have at your disposal a tremendous amount of forcefulness and a high level of physical energy. You can use these things to express and obtain your desires. And just what are your desires? Maybe you want to experience a leadership role of some sort, maybe you've never used your leadership abilities before; certainly you've never used them in the area you now find yourself using them in. Or maybe you've been upset about some condition you've seen in past lives and knew you should speak up about it but maybe you were afraid to, so now you're here to speak up about it or take action to do something about it.

Maybe your life last time around was totally centered in the emotional, the mental, or the material and now you want to focus more on the spiritual. (Aries, as a fire sign, is spiritual. earth signs are physical or material; air, mental; and water, emotional.) Unfinished business? Well, maybe impatience left things undone where Scorpio is. Or maybe there was some sort of destructive action here last time and now you have to repair or rebuild what's symbolized by this house. Above all else, this Mars is about manifesting. It's not enough to think about doing something; it's time to do it.

Mars in Taurus

You've learned to be a hard worker, to be dependable, and to sustain your efforts long enough to achieve what you want. So what is it you want? It might be material advancement or reward since Taurus is a physical or material sign. But it could be inner growth as well. Taurus has to do with maintaining the status quo. Maybe whatever you were doing here was so pleasant that you're not quite ready to let go of it, so you have to work at it until *you* grow out of it. You might have expressed a desire to test some sort of practical idea or to have an opportunity to take some sort of practical action in the area of business, agriculture, ecology, or some other "earthy" occupation. Or maybe you've been working terribly hard over many lifetimes and wanted an opportunity to put some energy into pampering yourself this time around. Unfinished business is apt to stem from excessive materialism, stubbornness, or opinionatedness where Scorpio is found.

Mars in Gemini

You're here in the hopes of experiencing more variety in your action-options. It may be that you spent your last life caught up in some sort of monotonous task that left you little time for anything else. Or maybe you were involved in some activity you enjoyed to the exclusion of all else and are now being forced to have more than this one activity going for you most of the time. Perhaps in your past life you over-controlled yourself, repressed your energy. Or perhaps you were in an environment or culture where it was very hard, if not impossible, to control your own life-direction because of your sex, religion, economic level, or some other factor. Maybe last time you failed at something you wanted to do and have come back because you feel a need to learn the proper way to do it. Gemini being a mutable sign, maybe you're trying to do two lifetimes of work in one this time for one reason or another. Unfinished business is apt to involve problems you didn't face or errors you refused to acknowledge last time around. Or maybe indecision kept you from accomplishing what you set out to do here.

Mars in Cancer

Either there's been a great deal of emotional turmoil in past lives and you've come back to sort things out and settle them down or you've felt unappreciated or unsuccessful. Either way, you've come back to get some sort of reassurance from others. You may be attached to domesticity or even to a particular locale, so you may have come back to continue being active in a parental, community, or even political capacity. You might particularly want children and may have incarnated to have them. Or you might feel a need to protect someone and have incarnated to take care of them as family members. You could even have incarnated to serve a country or to protect a country of your choice. Unfinished business is apt to stem from a past-life mixture of complaining and inertia. In other words, last time around, you complained about something but took no real action to change it. This time the "something" (or someone) is apt to be back again.

Mars in Leo

You might be here to take a more active role in the creative world, the business world, or the social world. Whichever it is, your desire is to gain more experience in this area. You may have felt, having done your apprenticeship in one of these areas, it's time to take action to become a leader. You may wish to strengthen your ability to get others to notice and serve you. Or you may just be thinking in terms of experiencing a bigger, better, fuller lifestyle. If there's unfinished business, it's apt to stem from past life self-centeredness—or at least from acting first and thinking later last time around.

Mars in Virgo

You've undoubtedly put a lot of energy into exploring the physical and material sides of life in the past; in fact, there may have been an over-emphasis on the physical—perhaps in the sexual area, perhaps in something like body-building or sports. Now you feel a need to either subdue your physical energies in this area or to make adjustment in your activity level in this area. You've undoubtedly developed very regular work habits and may be here to take on more responsibility in society, but your goal isn't to become president or anything similar. Rather, you're interested in using your energies in organizing or offering back-up support to another. Mars in Virgo often gives me the feeling that the person is at the crossroads of evolutionary experience in some way, at a point of just becoming aware of what life is all about and that there's more to life than one little corner of the world. Once this discovery is made, he or she has the option of retreating in fear or embracing the unknown. Often, with Mars in Virgo, feelings of inferiority threaten to hold the person back from realizing full potential this time around; whether or not there's any past life reason for these feelings of inferiority other than lack of experience in (or exposure to) certain areas, I'm not sure. If there's unfinished business in any case, it invariably stems from pettiness or from an inability to see the forest for the trees in a past life.

Mars in Libra

You've probably been busily working on your own in a relatively isolated situation on a fairly narrow range of activities. You've decided that you'd like a little more to life than what you've experienced, so this time you've asked for a wider, busier life-environment, which you have. And now you have to see if you can still function successfully without being overly influenced or distracted by others and without getting too many projects going simultaneously so that you wind up accomplishing nothing. Curiosity, too, may play a role in why you're back here at work. Perhaps you had to leave last time without knowing whether or not some problem you'd been involved in had worked itself out. Or perhaps indecision plays a role in your reason for being here; maybe you felt that you left last time without having enough time to weigh the pros and cons of a particular issue or course of action. Certainly indecision last time would play a role in any unfinished business this time, as could misguided gentility or backing out of something at the last minute last time around. Finally, you may be here out of a desire to enhance people or society in some way. In this case, you may have a desire to work as an artist or designer, a social worker, a surgeon, or even a hair stylist.

Mars in Scorpio

Unfinished business is certainly a theme, whether it's primarily heavy-duty karma caused by fear or vindictiveness or whether it's merely life-experiences you're still in the process of working on. Maybe you're here out of a desire to do something in an area few people know anything about. You may be here out of a desire to be an investigator or researcher in some little-understood area. Or you may be here to help some segment of society cope better; you could have chosen to be a psychiatrist, probation officer, or even an occultist. You may be here to intensify work or learning in some area that has been—up until now—a side issue. Or you may be here out of the realization that wishing isn't enough to make things happen, that you have to put physical, mental, and emotional energy into making them happen.

Mars in Sagittarius

This is a mutable fire sign, so there would be a tendency to try to do two lives in one or make up for lost time in some way. As a result, undoubtedly there will be a great deal of activity in this life, as well as some inclination to act without thinking. You might be here to experience being an authority (or teacher) in some specialized field; Sagittarius being a fire sign, this might be a spiritual field, but not necessarily. You might be here to develop your psychic abilities, particularly your ability to foresee events. Or you might be here to develop candor, honesty. If there's unfinished business, it's undoubtedly connected with verbal excesses of some sort in a past life. Usually these excesses are connected with having promised more than you could deliver and letting others down as a result. Another source of unfinished business could be the past-life tendency to make mountains out of molehills.

Mars in Capricorn

You're here on business this time around. By this I mean you're here to gain honor or status if you can. Your work is probably extremely important to you this time around (and if you're a homemaker, this applies to you as well—parenting and maintaining a home is definitely work). You've probably worked in this area over many lives and are now here to act as an authority and to claim the success you've been working toward for so long. You're here to see if you can succeed by using whatever is on hand rather than waste energy bemoaning what isn't here. Probably you're following a path you've tried before, hoping to get further along this path and preferably to the end where the rainbow is. If there's unfinished business where Scorpio is, it may have to do with past life pessimism that caused you to give up too soon. Or it may have to do with some sort of excessive earthiness—either sensuality or materialism—in this area.

Mars in Aquarius

You've developed a great deal of courage over past lives and are now here to use it—probably in fighting some counterproductive or stagnant tradition. You may be here to teach others how to save time. Or you may be part of a sort of cosmic emergency squad. It seems to me that people with Mars in Aquarius are often thrown into crisis situations of one sort or another; they have to do a great deal of work in a short period of time due to some threatening situation that involves themselves or those close to them. You may be here to influence or help past-life friends in some way. Or you may be here to see your past-life plans manifest on a larger scale. If there's unfinished business, it might stem from some sort of counterproductive attachment to the unusual last time around. Or it could stem from past-life worry.

Mars in Pisces

You're here because in the past you've done things you'd rather not be remembered for; you want to correct these situations so they'll be erased in your permanent records. You might be here to explore an artistic outlet, and in the process further develop your creativity. Or you might be here out of a desire to please others you've grown attached to. You might be here to develop your sympathetic qualities. Or you might be here to function as a healer on one level or another—not necessarily as a traditional healer or as a faith healer (though these are possible)—but simply as a caring, comforting person. You might be here to sharpen your psychic hearing or to learn to listen to your intuition. Unfinished business could stem from one of two very different sources—anti-social behavior or idealism combined with a lack of discrimination in the pursuit of your ideals.

These are, of necessity, very general guidelines. The full picture can be seen only by looking at your whole chart, so please think of these descriptions as starting points, not firm conclusions. In the remaining chapters, we'll add to this by looking at the house position and aspect position for Mars, which hopefully will allow you to synthesize and round out the picture.

Chapter 11

Esoteric House Positions

House positions for Mars will further clarify the options available to you and the goals you've set out to accomplish in life. In general, house positions for Mars (and to a lesser extent for the Sun, Moon, Mercury, and Venus) show circumstances you've chosen for yourself. Positions of Jupiter, Saturn, Uranus, Neptune, and Pluto show circumstances you've been more or less "guided" into, more out of necessity than out of a desire to grow. Planets within a degree of any house cusp have an aura of "cosmic coincidence" surrounding them. In other words, you didn't choose these options; nor were conditions forced on you. These are things that came about as flukes in the course of the determination of the "best" birth time for you. I've noted that people born by Caesarean, inducement, or born prematurely due to serious injury or illness to the mother often have more of these flukes than other people. People born before or after the given due date, but born without medical intervention to speed up (or retard) the birth, don't seem to have any higher incidence of flukes than people born more or less on schedule and without medical intervention designed to hurry things along. I caution you that this is a theory based on a sampling that's probably too small to be statistically valid, but for those of you who are research-minded, it bears looking at more closely.

And now for the obligatory word on house systems. If you're looking for the last word on the "best" house system, I'm afraid you're out of luck. What little research has been done on house systems doesn't point to a best one, although there are some indications that some house systems work better for some things than others. I played with a number of them before the first edition of this book was written, and I have played with a number of them—plus whole sign houses—since this book was written. My philosophy remains unchanged. What works for you works. Period. And if it works for you, then it's what you should be using, because it's the system that will give you what you need to understand the potentials of the chart, convey that understanding, and forecast important trends that an individual may find important to bear in mind.

In general, I see houses as akin to the lenses through which we view life. Those lenses are based on our particular philosophy of life and the particular environment and life experiences that have shaped us. So even though we're all seeing roughly the same thing, we may each see certain things more clearly than other things, and we may each put emphasis on different parts of the picture. What is clear to me in Placidus may not be clear to you in Koch until you convert it from Placidus to Koch. And while that may change the picture somewhat, it's likely that we are still seeing the same picture overall, albeit with a few differences in the details emphasized. We can only see what we can see from where we are standing. If we are missing things that others are seeing, then maybe we need to move to a vantage point where we can see more clearly. In the case of house systems, we probably want to look at things from a few different vantage points before settling on one spot, just in case there's a better view elsewhere.

When I wrote the first edition of this book, I was working with a tri-level system that consisted of looking at the chart in Koch, Equal, and Solar Houses, with solar houses showing the subconscious, equal houses showing the logical self, and Koch showing potential of the higher self. Over time I became less and less satisfied with this system and particularly dissatisfied with the results I was getting from Koch. I now primarily use Placidus—and I no longer see this level of being as specifically related to the higher self; I *do* see it as the level of manifestation and self-actualization potentially operating in someone's life. I still glance at solar houses, seeing these as perhaps an indication of the "heart's desire" or "spirit's desire" for this lifetime. And I see still see equal houses as sort of a case of "what is logical" or "all things being equal." Of late I have also looked at whole sign houses to see if they yield any better results, but at the moment the jury is out. Suffice it to say that if you prefer Koch to Placidus or whole sign to equal houses, by all means use them.

Esoteric Mars in the Houses

As this book is concerned primarily with Mars, only Mars placements are considered here. Positions can be read according to each of the systems above, with positions according to solar houses showing instinctive drives or actions, positions in equal houses (or possibly whole sign houses if you prefer) showing logical action or perhaps the inclinations based on thought, and Placidus showing the most likely manifestations of instincts and thought. Once again, if you find that Koch, or Porphyry, or some other system works better for you, feel free to use it.

Esoteric Mars in the First House

Any planet here symbolizes someone (or something or some action) that is extremely important in your life, for better or worse. Planets here encourage you to reach out, to seek experience. The more planets here, the more you'll want to experience in life. (Note that the Sun is always in the first house in a solar chart. There is, therefore, always a craving for some sort of activity or experience. Exactly what sort would be determined by the Sun sign to some extent, as the Sun sign has a great deal to do with what we are in the process of becoming.) If Mars is here, it's a favorable omen, as it says you've learned to approach life with a positive attitude. Your self-confidence is apt to be strong—or at the very least can be strengthened. The purpose of Mars here seems to be to give the energy and health necessary for you to attain your goals.

Esoteric Mars in the Second House

Planets in this house tell you what you need to possess and what you need to lose. They tell you both how you can make money and what you should do with it. If Mars is here, it may or may not imply involvement in a Mars-ruled profession. Mainly it gives you the ability to apply yourself to your work and do whatever is necessary in order to earn enough to live on. However, Mars here under stress from an outer planet would suggest that you'll have to be on the go a lot in order to satisfy your material needs and may have a harder time than other people in second house matters due to certain karmic debts or irresponsible past-life actions.

Esoteric Mars in the Third House

This house has to do with mental interests to be furthered and communication skills to be developed. If Mars is here in the solar third house, I take it as an indicator of mental immaturity. In times of stress, there will be a tendency to act without thinking because in the past communication has been more physical than mental. Otherwise, there will be an abundance of mental energy that can be used to further your goals in whatever manner you choose. Actions will involve learning right thinking, learning the consequences of too-impulsive thinking, and/or learning to compromise.

Esoteric Mars in the Fourth House

This house has to do with your home environment as well as your emotional and metaphysical development. If Mars is here, it suggests quite a strong emphasis on creative and/or psychic energy, which has developed over several lifetimes but hasn't been used constructively. In the past, impatience has gotten the better of you, and caused you to leave many things uncomplete or badly done. This tendency is still with you and will have to be watched closely. Your actions will involve finding a constructive outlet for your creativity and/or psychic ability—learning to develop thoroughness in your chosen field and learning to tone down strong emotions that interfere with your ability to follow your chosen path.

Esoteric Mars in the Fifth House

This is the house of creativity and self-expression. Like all of the succedent houses—the second, fifth, eighth, and eleventh—this is a regenerative house. Planets here push you to do something with what you have. "What you have" can be material or strictly potential. In any case, when you find planets in this house (or any of the other succedent houses) you're being told "use it or lose it." Of course, you can use it constructively or destructively, as you choose. If Mars is here, you're due for a lesson in self-control. There's a tendency to put all your energies into one area to the detriment of the rest of your life. Most often sex is the focal point, but less frequently it's a child or a business of your own that dominates your life. Selfishness has been a stumbling block for you in past lives; it can be again if you're not careful. Or you may pay your debts by experiencing a less-than-delightful relationship with a self-centered lover, bringing up a self-centered and difficult child, or finding obstacles in the career area, perhaps because those in a position to help you see you as a threat. There tends to be a perceived lack of freedom when Mars is here. My guess is that you abused or refused freedom in the past and are now paying the piper. Not fun, but unless there's a slow-moving planet here, you chose this path of your own volition and undoubtedly you're capable of gaining from it.

Esoteric Mars in the Sixth House

This is the house of health and improvement. Planets in the sixth house symbolize things we haven't done well—misguided efforts. Chances are your intentions were good but your ideas were flawed. Or perhaps you chose to attempt things that were for one reason or another beyond your capabilities. If Mars is here, there is a tendency to try to push your way through life, to try to make up for lost time in the past. This could also mean that your karma involves a great deal of physical work or work involving machinery. Or it could involve psychology. There would be a tendency to run high fevers when ill. Illnesses may or may not be of karmic origin, depending on your overall chart.

Esoteric Mars in the Seventh House

The theme of the seventh house is "we vs. me." It's most specifically involved with marriage, but business partnerships and relationships with clients are also a part of this house. If you're an astrologer, counselor, hair stylist, et cetera, pay special attention to this house; it will tell you a lot about why you're doing what you're doing. If Mars is here, there's been impulsiveness in the area of marriage in the past. This tendency would have to be curbed now, and any past damage done would have to be rectified. Karma might be worked out through experiencing a marriage to an argumentative person and in extreme cases your mate might even be physically or mentally cruel. (However, I have also seen cases where, while there was no cruelty, there was an extremely dependent relationship, which, while warm and comfortable, also interfered with the individual's growth as well as the growth of the significant other.) In fact, sometimes you alternated between these two extremes of smother love and animosity and it's now time to get it straight once and for all.

I will digress here to give my well-known and not-too-well-loved sermon on soulmates—or at least the part of it that's applicable here. The concept of soulmates is much misunderstood and has become a favorite excuse for wishywashyness and just plain laziness in the marital area.

Lest you think I'm saying something I'm not in my discussion of the seventh house where Mars is concerned, I'll state the facts as I know them:

1. There are soulmates.

2. If someone is truly a soulmate, you'll be together come what may.

3. There's a very good chance that in order to be together you'll have to overcome obstacles, put up with delays, or make some sort of change in your philosophy of life, lifestyle, or goals. BUT:

4. When you discover your soulmate is someone else's mate, or you discover your soulmate after you've already married someone else, you'd better carefully examine your motives and your concept of karma. Granted, occasionally you'll actually meet your soulmate at a time when one of you isn't free. This normally has a lot less to do with karma than it does with your (or his or her) own impulsiveness or faulty concept of what marriage is all about. Often, the person you meet isn't a soulmate at all, but is rather a cop-out solution to boredom or marital problems you're too lazy to work at solving. In this case, even if you do actually marry, once the honeymoon is over you're liable to find that you're no more content than you were before.

5. If you're truly soulmates, you'll both realize it and take appropriate action, because you'll know that's the only thing that can be done. A soulmate won't give you lines about staying married for the sake of the kids, a suicidal spouse, potential financial ruin or anything else. He or she will do what needs to be done. Yes, there may be sacrifices involved, and certainly these things take time, but if it's been four years and one of you is still procrastinating, this isn't a soulmate relationship; it's a case of greed, of wanting it all. Getting it all isn't what karma is all about; karma is about learning to fulfill needs and choose priorities wisely.

6. Finally, and most important, karma does not require you to spend your life being battered, sworn at, or being treated for the STD your spouse brought you from his or her latest trip out of town. True, you may have in the past been less than an ideal mate. You may have been a drunk, a bully, or an adulterer. You may even have to learn the hard way that it's no fun to be on the receiving end of this kind of behavior. But you aren't required to continue experiencing it once you've realized what it feels like. If your mate refuses treatment and you refuse to leave, this isn't paying a debt; it's refusing to accept responsibility for yourself and your actions (or inactions, as the case may be). This is voluntarily choosing to waste your life and perhaps doing harm to others in the process. Soulmates come together to help one another, not destroy one another.

And that, dear readers, is what tends to be conveniently forgotten in the heat of passion, the quest for excitement, or the desire to let someone else take responsibility for your life. End of sermon!

Esoteric Mars in the Eighth House

The eighth house is sometimes called the house of death, sometimes called the house of transformation. It's another of the "'use it or lose it" houses. Planets here demonstrate the occult principle that thoughts are things. If you have planets here, be careful what you wish for. You just might get it, and if you do, it'll be mighty hard to get rid of if it's not to your liking. If Mars is here in the Equal House system, it could be a sign that you've misused your psychic powers in a past life and now must pay the piper. This can mean there will be almost an obsession with obtaining some sort of psychic information or becoming skilled in a metaphysical discipline, or it could mean there will be a strong fear or dislike of the occult and its practitioners. Either way, there will be a blockage of your psychic powers—you'll be able to go so far and no further. As a routine precaution, you should stay away from intense quasi-religious groups, avoid techniques that require you to go into trance, and take advice from nontraditional healers with a grain of salt. If Mars is in the Placidus system, you're at the "use it" end of the spectrum and therefore may well become active in metaphysical groups or in one or more occult disciplines. (Alchemy, magic, and esoteric astrology seem to hold particular appeal for those with this placement; on another level, so does psychology, especially the more humanistic modes of psychology.) In either case, there's a good chance that others will seek your advice and this is how you'll work out your debts.

Esoteric Mars in the Ninth House

This is the house of the higher mind. Planets here are given to increase your understanding of life in general. They also influence your personal code of ethics by expanding and/or refining it. If Mars is here in the Equal House system, you've had a very dogmatic philosophy of life in the past. You tended to see people whose beliefs differed from yours as bad—even dangerous—people. You could have participated in the Inquisition, the Salem Witch Trials, or some other nasty piece of history, or you could have been just a run-of-the-mill Archie Bunker type bigot as a result of a combination of ignorance and self-satisfaction. If Mars is here in the Koch system, you're here to be a pioneer or ground-breaker for some school of thought that departs from tradition. This may be in order to pay a debt caused by past closed-mindedness, or it may have something to do with developing the courage of your convictions. The chief difference between the two Mars positions is that the equal house position stresses that you open your own mind, while the Placidus house position involves opening other people's minds. (Of course if Mars is in the ninth house in both systems, you're to do both!) Regardless of house system, this position tends to give a very restless mind, and, of course, learning to control this restlessness is one of your reasons for being here. There may also be a desire to work in a foreign country and/or teach and learn about a foreign country's people, customs, and philosophy. Doing one or all of these things may be instrumental in working out your karma.

Esoteric Mars in the Tenth House

This house shows your working conditions and your status on the material plane. Planets here show you what you did to deserve your lot in life. If you hate your job and can't get "no respect" here's why—and what you can do about it. If Mars is here, you may work out your karma in the field of mechanics, psychology, or technology (especially technology in which machines are used to replace people for tedious chores). Actions would involve making your desires realities, and of course Saturn would test you at many points along the way.

Esoteric Mars in the Eleventh House

This is called the house of hopes and goals. Planets here represent plans that have not yet materialized. What you do in this life will either change these plans, perhaps even cause them to be abandoned, or bring them closer to fruition. If Mars is here in the solar eleventh house, you can be strongly drawn—almost compelled—to act as a leader. How successful you'll be in this role depends on your aspects to Mars. Suffice to say that you've probably played this role or a similar one before (unless Mars is in Aries, in which case you're fed up with being a follower and this is what motivates you to lead). If Mars is here using the Equal House system you may be attracted to work connected with machinery, science, or psychology as a means of working out your karma. If Mars is here in the Placidus system, you might be involved in developing machines that will do work for people. Or you might be involved in retraining people who've been displaced by machines or helping people discover constructive, enjoyable uses for their leisure time. Regardless of house system, Mars says that actions will involve working for the betterment of groups of people or society as a whole. This may involve participation in some formal association (which could be either large or small) or it could be a more informal effort that involves you and a friend or two. In any case, this work would help pay a debt incurred because of your selfishness in the past. Actions may also involve learning when, why, and how to say No! as chances are good that in the past you've had problems in this area that have kept you from growing.

Esoteric Mars in the Twelfth House

This is called the house of sorrows or the house of self-undoing. It can fit that description—but it doesn't have to. This is the house of your subconscious. And if you descend to the depths of your subconscious and try to understand what you find there, you'll transcend your sorrows and discover your soul. Actually, the gloom and doom connotations of this house are, if not total misconceptions, at least highly over-rated. Planets here symbolize those things that separate you from the mainstream of life. These things may be traumas, physical limitations, inferiority feelings or any other grisly thing you care to imagine. They may be self-inflicted or acts of God. Either way, you're responsible for them in the sense that it's you—and nobody else—who allows them to make you unhappy instead of letting go of them or learning to work around them or turning them into strengths rather than weaknesses. Which brings me to another point: amidst all that psychic garbage, there are other things that separate you from the mainstream of life. These are your inner strengths, the results of lessons learned, knowledge gained, and obstacles overcome in your past lives. They may be talents such as writing, creating, or a special ability to understand others. They may be character strengths such as courage, practicality, or self-awareness. They may be guardian angels, either in the sense of intuitive or psychic abilities, or they may be people to whom you've given help in the past, people who will now help you if you open yourself to their caring. In any case, these things exist. For everyone. The more planets you have in the twelfth, the more strengths you've developed. But to tap these inner resources, you first have to wade through all sorts of soggy flotsam and jetsam and get rid of your "poor me's" and "life's unfair's" to find and understand the real you. If Mars is here, temptations are the theme. With the Equal House position, the temptations tend to be tangible or material. With the Placidus house position they tend more towards self-deception or self-defeating behavior. There's a tendency to repress your feelings and become quite passive unless Mars is conjunct the Ascendant. This must be watched, as it's the prime cause of self-undoing with this position. Your actions will involve developing confidence, acting on your convictions, and overcoming resentments.

Remember that if in any position a planet is within a degree of a house cusp (assuming, of course, you know your birth time or have had your chart rectified), it may be a fluke and what has been written may not apply to you in terms of karma.

Chapter 12

Esoteric Aspects

Aspects can be used to further refine your view of your past in relation to your present. Any aspect used in traditional astrology also has meaning in esoteric astrology. However, for the sake of brevity, we will cover only the conjunction, sextile, square, trine, and opposition, and these only in relation to Mars.

Conjunctions symbolize potentials to be developed or energies to be synthesized. They emphasize certain feelings, talents, or environmental factors and in that way push you to do what needs to be done. The conjunction says "make the most of it," or "get your act together." Conjunctions may operate on an inner level, an outer level, or both.

Sextiles show opportunities for progress or success. They symbolize chances to return favors given in past lives, chances to benefit from good work done in past lives, and chances to discover new people, ideas, and things before you're actually forced to deal with them. They are, in other words, evolutionary short-cuts, if you use them consciously. They don't give you any free rides. If you choose not to use them, they'll just sit there nicely and not bother you. But if you have them, you'd be foolish not to use them. Most people can use all the help they can get and sextiles, when used, definitely fall into the helpful category.

Trines show you what you've successfully developed or dealt with in the past. Therefore, trines symbolize things that come easily to you. They protect you from situations, relationships, and ideas which are as yet too complicated for you to successfully deal with; they don't, however, protect you from your own laziness should you choose not to use what you've learned to date. Unlike the sextile, the trine shows benefits that will come to you without much effort. They symbolize other people's debts to you that have now come due. In most cases, they'll be paid promptly and willingly; in a few cases, they may be paid more grudgingly, for a planet may be in trine to one planet and square or opposing another.

Squares have a bad reputation. They show you things you haven't yet mastered and therefore have to do with struggle, depression, being misunderstood, and generally finding out that you can't always get what you want. Yet, squares are very necessary. A person with no squares has no motivation to grow; this person may have fun, but he or she won't learn very much, and probably won't accomplish very much either. A person with lots of squares will have more power than someone with a lot of trines, and will accomplish a great deal in life if this power is used. That's what squares are all about—learning to handle power (and, conversely, learning to accept limitations). Squares symbolize surmountable obstacles. These may be environmental or inner. Either way, they affect you in a material, tangible way, as they manifest and are worked out in the course of your day-to-day activities. No amount of contemplation will eradicate them; you must meet them head on and deal with them in the real world.

Oppositions symbolize energies or goals that must be reconciled. They're easier to handle than squares in that with squares if you want to deal with the problem symbolized, you have to do it yourself. Nobody is going to help you figure out what's wrong and nobody is going to help you rectify it either. Squares are problems that can only be solved by you. Oppositions, on the other hand, bring other people into the picture. These people will tell you exactly what is wrong and will help you rectify it. Indeed, they'll probably force you to take steps to rectify it! Squares can, if you're self-deceptive, be passed off as "just your imagination" or other people's hang-ups that unfortunately cramp your style but don't really have anything to do with your own failings. (People who don't work with their squares frequently accuse other people of taking advantage of their generosity, being jealous of their talents, or being intimidated by their intelligence, generosity, talent, and intelligence of course being nice things—rather than curbing personal wishywashyness, arrogance, or showoffishness; which, unfortunately are the flip sides of the nicer qualities.) Squares you can live with. You may not be totally content, you certainly won't reach your fullest potential, but if you're willing to settle for crumbs, you can just let your squares sit there. With oppositions you don't have that luxury, because when you ignore them you'll become so uncomfortable and unhappy—or people will become so unpleasant and disappointing—that your survival and/or your sanity is at stake. My feeling is that if you ignore your squares over several lives they eventually become oppositions. I therefore feel that oppositions symbolize major debts that are long past due. As such, they symbolize things that can't as a rule be fully rectified in one lifetime—unless you have a Grand Cross in your chart. In that case, you've contracted to work overtime this life and complete two or more "lessons" at a sitting. So, while oppositions may seem easier to handle on the surface (because of the help available to you in terms of identifying and even solving the problem) they're really indicative of deep-seated bad habits, while squares are merely potential bad habits that can be nipped in the bud and turned into plusses. Oppositions, then, can't be ignored.

Conjunctions and Harmonious Aspects

This first group of aspects is conjunctions and harmonious factors (sextiles and trines), and I have differentiated where necessary between the sextiles and trines.

Esoteric Sun-Mars Conjunction

Perhaps you misdirected your ambition in the past. Perhaps you were unnecessarily aggressive or you refused to compromise when you should have. Or perhaps you intentionally (or through your own carelessness) caused harm to someone else. You probably haven't really been stupendously bad, just selfish or insensitive. As a result you may find yourself having to deal with erratic, aggressive people, or maybe you'll have to take care of someone who is in some sense weak and incapable of looking after himself or herself. Or maybe you'll be accident-prone. Your power is simply power for doing. Doing what and for whom can only be seen from your chart.

Esoteric Moon/Mars Conjunction

There's been excessive passivity and an over-emphasis on logic in the past, so this time feelings will dominate. You may be hot-tempered and act impulsively as a result. The exception would be if Saturn is square, trine, or opposition either planet. In that case, you mastered your emotions once and are now being tested to see if the lesson "took." You therefore have better self-control than those without the Saturn aspect but will still have to watch yourself to make sure you're neither repressing nor over-reacting. In any case, the house position of this conjunction would be very important in terms of showing you both constructive and destructive emotional outlets.

Esoteric Mercury-Mars Conjunction

You learned in the past to be very independent in your thinking. So now you've earned the right to be recognized for some method you devise or something you do. The sign the conjunction is found in, as well as the other aspects to Mercury and Mars, are very important in terms of giving you the full picture as these show your aptitudes and most comfortable speed of working, and provide you with some idea of how to use your mind and energies to the best advantage.

Esoteric Venus-Mars Conjunction

There could have been excessive sensuality, adultery, or greed in the past. There is tremendous ambition for a romantic/marital relationship because in the past romance and marriage were less than satisfactory. You may have been excessively aggressive in the past, so one of the reasons you're here is to learn to be less pushy.

Esoteric Mars-Jupiter Conjunction

You were a teacher of some sort in the past, so you'll enjoy introducing other people to your interests this time around. You've been given plenty of energy—perhaps more than you need. There's also a possibility of gain resulting from efforts made in the past.

Esoteric Mars-Saturn Conjunction

This is given to teach you patience. You've been very practical in your actions in the past but have perhaps given up too easily. This time, you'll let nothing stop you.

Esoteric Mars-Uranus Conjunction

In the past, you had a violent temper. A tendency toward hot-headedness is still present, so your mission is to learn to control it. You also need to work at gaining other people's cooperation; before, you antagonized people and therefore couldn't complete all of what you set out to do. A need to create things may be part of your karma, as creativity may have been blocked or remained undiscovered in the past. Lack of concentration has almost certainly been a problem before, so this will have to be worked on. You'll tend to be high strung in this life and will have to consciously work on these lessons if you don't want to have to learn them the hard way.

Esoteric Mars-Neptune Conjunction

If one or both planets are in the third decanate, there's been involvement with mysticism or occultism in the past. In any case, you'll need time to be alone because one of the reasons you're here is to further develop your intuitive abilities. I feel that one purpose of this conjunction is to give you a rest from the limelight (or at least from the nine-to-five rat race). For this reason, people with this conjunction are generally much happier when they can work alone and at their own pace.

Esoteric Mars-Pluto Conjunction

You're here to accomplish something very difficult that was probably started long ago. Sign and house positions will give you further clues.

Esoteric Mars-Ascendant Conjunction

There's been too much dependence in the past. This placement is given to help you develop self-reliance.

Esoteric Sun-Mars Sextile or Trine

You've been denied a promotion or an increase in status in the past, and this has deeply disappointed you. So now you'll work hard, but your efforts will continue only as long as you feel there's something to be gained by them. In other words, rewards are important to you. The sextile gives you an opportunity to learn about right effort and right motives.

Esoteric Moon-Mars Sextile or Trine

Illness or low energy might have kept you from accomplishing your goals in the past. These aspects have therefore been given to encourage good health and a good energy flow. They also increase the potential for romantic opportunities.

Esoteric Mercury-Mars Sextile or Trine

There's been crudeness and/or ignorance in the past, much to your regret. Now you've been given above-average intelligence, although it will be up to you whether or not you use it. Crudeness and impatience are still present, so you'll have to watch yourself when you lose your temper as you can say things you later regret. Crudeness tends to be more common with the sextile than with the trine, although if Venus is conjunct, sextile, or quintile Mercury, this tendency would be offset.

Esoteric Venus-Mars Sextile or Trine

You've asked for an opportunity to pursue a creative interest and therefore could be very active in an artistic area or in something like cooking or fashion. The sextile helps you attract both financial and romantic opportunities.

Esoteric Mars-Jupiter Sextile or Trine

You've come back to be a worker rather than a nobleperson or person of inherited wealth. Your actions will therefore involve direct experiences more than theories. The sextile gives you an opportunity to reform someone or something.

Esoteric Mars-Saturn Sextile or Trine

You were a homemaker in the past. While probably relatively happy, you disliked having to be dependent on your mate and may have felt that your work was insignificant. You therefore will feel a need to work outside your home in order to have a sense of accomplishment and to know how much (in material terms) your skills are worth to others. Security will continue to be important to you, but this time you'll want to have a more active role in providing this for yourself and your family. The sextile will give you opportunities to develop your courage in some way.

Esoteric Mars-Uranus Sextile or Trine

Any Mars-Uranus aspect can indicate genius or at least a high level of creativity. Sextiles and trines, however, don't insist on being used, so while you may find mental work easy, there's no guarantee that you'll do anything extraordinarily noteworthy. You could, but no one will force you. Over your past lives you've developed a great deal of resourcefulness. You're now here to learn to be more positive in your efforts and actions. There's no difference between the sextile and the trine in this case.

Esoteric Mars-Neptune Sextile or Trine

You're a sensitive person and have been involved in work that served humanity in some way in the past. You'll therefore be sensitive to the needs of those who aren't able to take care of themselves for one reason or another, and you may be instrumental in helping these people get what they need or become more self-sufficient. The sextile gives you opportunities to be protected by your intuition or by unseen powers.

Esoteric Mars-Pluto Sextile or Trine

Your vitality has been poor in the past. Maybe poor health handicapped you; in any case, your lack of accomplishment wasn't due to laziness and was regretted by you. So now you've been given the ability to survive on very little sleep and recover your strength very quickly after working hard for extended periods. This is a test. You can, therefore, choose to use your energies for drinking, sex, and rowdy partying or for something constructive. There's very little difference between the sextile and the trine.

Esoteric Mars-Ascendant Sextile or Trine

You've been too impulsive for your own good in the past. There's still a tendency to be a little too impulsive, but this time you'll have opportunities to learn how to control this trait.

Squares and Oppositions

The section that follows outlines discordant aspects (squares and oppositions) so you can see how Mars expresses itself esoterically. These aspects are important to understand for maximum growth in this lifetime.

Esoteric Sun-Mars Square or Opposition

These aspects are given to teach you patience. The square has been accused of being a draining aspect, but I don't find this to be the case. It gives more physical energy than any other Sun-Mars aspect except the conjunction—certainly more than the sextile and trine. But energy isn't used as efficiently. It's more liable to be used destructively at the outset, which means you must then use still more energy to get yourself out of whatever mess you got yourself into in the first place. This, not low vitality, is what causes people with this aspect to feel drained. The square symbolizes smugness and/or excessive egotism in the past. If these are retained in the present, your actions will be self-defeating rather than constructive. By being in a hurry to prove a relatively insignificant point, you'll do things that make your intentions suspect and actually take you further from, rather than closer to, your goal. By slowing down and not worrying so much about whether or not others give you an instant agreement, you'll succeed in using your energy constructively.

The opposition most often pertains to sexual lessons and is, I think, an indication of having many more male incarnations than female ones. This has caused you to have an imbalance between the active, outgoing, masculine side of your personality and the passive, reflective, feminine side. Regardless of which side is over-emphasized, the end result is discordant relationships with men. When you learn to keep the active and passive sides of your personality balanced and at ease with one another within you, the problems in your outer life cease. However, as this imbalance has developed over several lifetimes. you'll have to consciously and consistently work at striking this sort of inner balance throughout your life or problems will recur.

Esoteric Moon-Mars Square or Opposition

Creative talents are strong—stronger than with the sextile or trine—but there's a tendency to feel these are useless. This feeling comes from past-life experiences where either your talents were belittled and unappreciated or else you took on some sort of stupendous project that virtually doomed you to failure from the start. You need to experience a sense of accomplishment, so regardless of which aspect you have, you're better off working on a series of short-range projects than on one highly detailed long-term project. Why? Because one accomplishment will lead to another and success will breed success.

The square suggests that domestic harmony will be absent at one point in your life, in part due to your own feelings of inadequacy. Once you realize your own worth and discover your own strengths, this situation will improve. The opposition is also indicative of domestic discord, but in this case the situation is more complex because it's a continuation of some past-life discord. Problems can be overcome, if not alleviated, by developing emotional self-restraint and channeling excess energy into some sort of creative outlet.

Esoteric Mercury/Mars Square or Opposition

You're great at starting things, not so great at finishing them. Or maybe you're more talk than action. My theory is that these aspects are given to jolt you out of past-life lazy-mindedness.

The square suggests dishonesty in the past. You have to be on guard for this tendency getting the better of you in the present as well. This past-life dishonesty was of a minor nature and hurt no one but you. The opposition is given to strengthen your determination. You've been a yes-person before. This has hurt you, and has also misled or held back others. Now these others are back to challenge you.

Esoteric Venus-Mars Square or Opposition

You've been in a position that can best be described as downtrodden. You had no respect from your mate, your work efforts weren't adequately rewarded, and you've been forced to shut up and take orders whether or not you felt they made sense. This left you feeling crushed and resentful. You've carried your sensitivity from the past into this life, but you've also carried in open wounds. So this time around, you'll fight back—in fact, at times you may become very obnoxious. You'll want to be a leader and will demand respect from your mate, your boss, and anyone else you deal with on a regular basis. If you don't get respect and recognition, you'll see to it that somebody is sorry. The problem is that you'll be well able to voice your complaints and criticisms, but if others so much as suggest you might be less than perfect you'll be terribly hurt and, rather than trying to figure out if their complaints are valid, you'll automatically feel you're (again) being mistreated.

The square shows a tendency to live for today. When you want something, you want it as soon as possible. This tendency has you in a constant state of tension that generally hurts nobody but you in the long run. Occasionally this aspect suggests promiscuity in a past life. Again, this hurt nobody but you (although there may have been an abortion involved). In this case, there might be either romantic difficulties or problems with your reproductive system. The opposition suggests that your lack of assertiveness, or your promiscuity, has harmed others as well as you even though you may have been more victim than villain in the past. Certain people will be back and you'll have to work out your difficulties with them. The problem is that at times you'll be tempted to be disruptively aggressive, which of course will only compound the problem. Use care when transiting Mars, Neptune, or Pluto makes a conjunction, square, or quincunx to either natal Venus or natal Mars. If you can keep yourself from making outrageous demands in these periods, you'll work out your karma much more constructively than otherwise.

Esoteric Mars-Jupiter Square or Opposition

You lost your individuality, your true concept of self, through over-identification in the past with a group, cause, or another person. Undoubtedly you were a good person, but you were somehow misguided. So now you have to strengthen your identification with your true self. The square suggests a lot of wasted time in the past. This tendency, along with a tendency to become easily distracted by superficialities, must be watched in the present.

The opposition suggests a neglect of health in the past, or perhaps some sort of misguided self-denial to further a cause—maybe starving yourself to death, self-immolation, self-mutilation—something dramatic that really didn't do much for either you or your cause. As a result there could be some sort of blood disorder or an accumulation of toxic materials in the system this time around. The opposition can also be saying that your involvement with a cause, group, or individual in the past caused you to neglect other facets of your life and this somehow burdened other people unnecessarily. In this case, you can be pretty sure that other people are going to keep you busy with all sorts of work that needs doing that (they say) nobody else has time for.

Esoteric Mars-Saturn Square or Opposition

You're capable of working very hard. You've developed a strong sense of discipline that can be very helpful to you in terms of helping you accomplish what you want to do. But last time around you overstepped your authority or used force where you could have used reason. This time around if you use similar tactics, the powers that be are likely to curtail your freedom.

The square suggests a restriction taken on voluntarily in order to pay a debt. The nature of the restriction will depend to a large extent on the signs and houses involved. The opposition suggests an involuntary restriction placed on you for your own good. You should concentrate on the house Saturn is in. This house represents a person you're indebted to. He or she will test you; if you've learned to use your authority wisely the restriction will be lifted to some extent. If not, this person is going to give you a pretty rough time.

Esoteric Mars-Uranus Square or Opposition

You've developed a great deal of willpower in the past. Your patience, however, has been sadly lacking. Now this must be rectified.

The square says you've gotten in a rut, gotten so comfortable with certain modes of behavior that you haven't explored or developed other potentials inherent in you. As a result, there are apt to be abrupt changes in your life from time to time. In extreme cases, maybe even an illness or injury forces you to reexamine something you've taken for granted, or it forces you to develop certain talents that might have otherwise remained latent. The houses involved will show you the areas of your life most likely to be affected by change. In the case of illness, the signs involved will give you further information. The opposition says your bad temper and impatience in the past led to rather serious relationship problems. Expect to have someone with a quick and fiery temper around you in this life, and expect to have to compromise with this person if you don't want problems.

Esoteric Mars-Neptune Square or Opposition

You've been deceived by others in the past, but this is primarily because you let yourself be deceived. In other words, you've been your own worst enemy. The square suggests mild over-indulgence or sexual excess that hurts no one but you. You probably spent a lot of time flitting from party to party last time. The house containing Neptune will tell you about things that can deceive you or waste your time this time around, while any house or houses with Pisces on the cusp will show you where confusion, gullibility, or lack of self-awareness can cause you to deceive yourself.

With the opposition, you have to be careful who you accept help from as the people you come in contact with may not have the purest of motives. They may help you in the hopes of making you beholden to them so that you'll do what they want rather than what you need to do. In general, people will promise more than they deliver.

Both of these aspects suggest that there's been some sort of involuntary psychism in the past. Usually this is involuntary mediumship, clairvoyance, or astral projection. It may have been used; it may have been repressed and come out as eccentricity, epilepsy, or mental illness. It may have led to you being misused by an unscrupulous person for his or her gain. In any case, I usually recommend a good introduction to psychic phenomena when I see either of these aspects, as the ability is still present, and if it comes out involuntarily it can be frightening and destructive. Used properly, however, it can be a great asset, since it will help you steer clear of deceiving, self-serving people. You would, however, have to shop very carefully for your source of information, regardless of whether you choose a teacher or self-study through books, since until your intuitive powers are firmly in your control, the tendency towards self-deception and/or deception by others remains.

Esoteric Mars-Pluto Square or Opposition

You've been violent or self-destructive in the past. The square says you harmed only yourself; the opposition says other people were also harmed. You must now be very careful to avoid violence, excesses, and fanaticism. The involved houses will give you further information.

Esoteric Mars-Ascendant Square or Opposition

You've been too forceful for your own good in the past. You must now strive to win through co-operation rather than coercion. With the opposition, don't be surprised if you meet with more than your share of hot-tempered, bossy people who rush you or create problems for you through their carelessness.

Chapter 13

Sample Mars Analysis

Let's look at a sample chart to see how the Mars energy manifests itself for a real person. The chart on page 156 (Koch House System at the request of the client) belongs to Rose and we're going to see if Rose's Mars tallies with the information she has volunteered about herself. We know from talking with her that she is married for the second time, and has four children, one of whom is a teenager still living at home. Though not presently working outside her home, she has been a singer, a real estate agent, and a district manager for a large door-to-door cosmetic sales firm. Her current primary interests are home and family, holistic health, and metaphysics. She does a great deal of volunteer work and is proficient at a large number of handcrafts.

Rose has Mars in Leo, so we'd expect a fairly high level of physical energy and a more or less even output of energy. This energy would be expressed confidently. A lot of it would be channeled into getting others to do what she wants them to do. Someone who has performed in musicals and supper clubs must, of necessity, project confidence, and of course success in sales involves getting others to do what she wants them to do. She definitely has ability to inspire and motivate people, as her sales and volunteer work records attest.

Rose entertains frequently and well and is described by friends and colleagues as a warm and generous person. Is she able to work without recognition? Yes, but she prefers not to! And she isn't shy about taking credit due her. There's no false modesty here!

Mars by sign and house is an indication of what makes people mad. In Rose's case we would expect her to get mad if she's taken for granted or ordered about (Leo) or as a result of domestic glitches (fourth house). With a fixed-fire Mars, we'd expect a long fuse on the temper with occasional bouts of rather fiery pique. Contentious issues, once discussed, would probably not be re-

156/The Mars Book

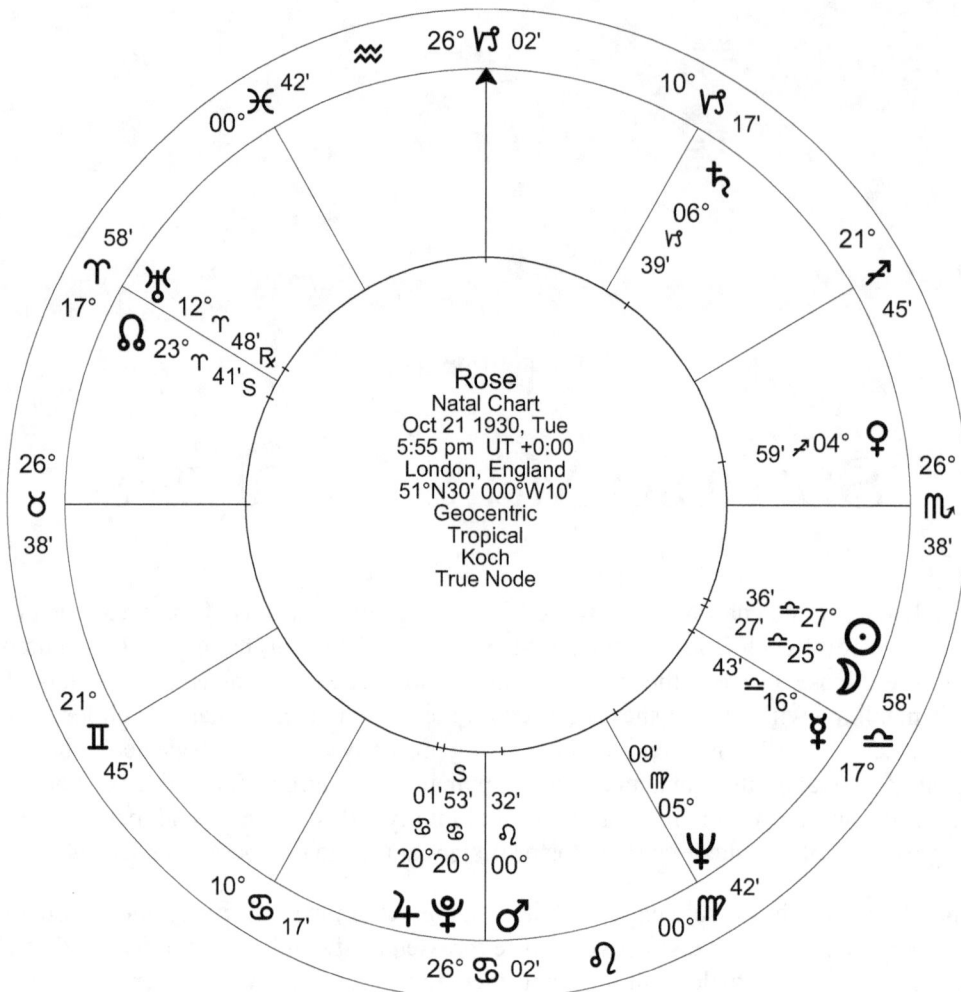

hashed; however, they wouldn't be erased from the memory banks either. Rather, she'd take a wait-and-see approach until the problem came up again and would then act accordingly. As you would expect, Rose is a hard worker in her home, which is well maintained and attractively decorated. In the time that I've dealt with her, most of Rose's exasperations have been connected with what I would call other people's lack of respect for her home or by tradespeople who have not met deadlines.

Rose's fourth house Mars suggests a somewhat strife-ridden early life, and in some respects this is true. Although Rose did not lose a parent in childhood, she was separated from her parents during the war for a time because her hometown was under attack. As was the case with many children, Rose was evacuated and spent a year in a "safer" part of the country where she and a friend worked as household helpers. Air raid and bomb shelters were a reality of Rose's teenage years. (Mars in the fourth, as well as Pluto in the fourth, are themes I have found frequently in charts of people affected more or less directly by war.) Other types of stress affected Rose's early years as well. There was an older brother who occasionally took his frustration out on her by hitting her. There was also a mother who was, as we say in the trade, "a strong personality."

I have found that there are no hard-and-fast guidelines as to which house rules which parent. In Rose's case, Cancer on the fourth led me to suspect the fourth might be Momma. Mother was described as the person who gave the orders and ruled the roost—bossy and perhaps a bit judgmental. In contrast, Poppa was described as a rather quiet man. There was strife with Momma, where Father was largely respected. Does it sound like Momma might be fourth house? It does to me, although I suppose this could be argued. While there was no divorce and no active violence in Rose's home, there was what I would call passive-aggressive discord. In other words, Rose's Momma steamed, stormed, and slammed doors, while Poppa remained relatively stoic and dignified.

Rose has what I call a square with safety edges—an out-of-sign square—between the Sun and Mars. In my experience, these are somewhat less angst-producing than "regular" squares, though they can still be indicative of problem areas. In Rose's case, we might expect some home/work conflict (fourth/sixth) or some conflict between her need for harmony (Libra) and her need to assert herself. Since both Leo and Libra are signs that need praise, the bottom line with Rose is probably "getting things done right"—as per her concept of "right," of course! In other words, she will be aggressive at the outset in order to prevent rejection or criticism in the long term. Another part of this process of getting things right for Rose is the study of holistic health, which is encouraged by both the Sun-Mars square and the sixth house Sun-Moon conjunction.

The Moon-Mars square also has "safety edges." This can be a somewhat defensive, contentious aspect, and could create ambivalence as to whether work achievements or family-related achievements will take priority. It could also indicate a low tolerance for stress or discord and a tendency for these to have a deleterious impact on her health. On a behavioral level, this aspect could indicate "leap before you look," accident-prone behavior or "imploding emotions"—imploding rather than exploding because Mars is in a fixed sign (denoting a long fuse on the temper) and Moon in Libra is brought up to be nice at any price. Additionally, this aspect could indicate a likelihood of working or living with difficult people who manifest the defensive, contentious side of this aspect.

With Mars in the fourth square both Sun and Moon, we have good reason to suspect discord between Momma and Poppa. There *was* discord, but it was that peculiarly Libran sort of discord where Momma ranted and raved and Poppa said, "Oh, well, that's just the way she is." True to her Mars in Leo, that wasn't good enough reason for her to sit back and accept it along with Dad.

Rose *hated* her mother in her early years, though in true Libran fashion she made her peace with her in later years and grew to be objective about her.

The stress of the war years, working at age thirteen (when she started singing in supper clubs), the "cold war" at home, and undoubtedly puberty, all took their toll on Rose. At age sixteen, she had a nervous breakdown, triggered by a failed infatuation. She was hospitalized briefly—an experience she describes as a positive step on the path to self-discovery and acceptance of her environment and her potential.

If these two squares have safety-edges, then the out-of-sign sextile between Mars and her Ascendant might be described as a chipped sextile. Think of a rough-edged fingernail—perfectly usable and no bother for the most part, but it can catch in delicate fabrics and cause snags. This aspect suggests that Rose wants to be more or less in charge of her own life and will have many opportunities to determine her own course of action. With Mars in Leo, she doesn't want to be taken for granted. Nor would she like feeling obligated. She doesn't want a lot of demands placed upon her without her okay; this is a lady you ask—rather than tell—to do things. And yet, Mars in the fourth house has something of a need to be needed. Rose very much needs the option of accepting or refusing duties. If she feels she's being drawn into a situation that involves freedom-curtailing expectations, her tendency would be to instinctively withdraw from it, possibly by taking on an even heavier responsibility of her own choosing or pleading family obligations. If, however, these responsibilities are taken so seriously by others that they cease to call and offer options for involvement, Rose's ego may get a bit bruised. In other words, in an attempt to protect herself, she may occasionally cut off her nose to spite her face, either by opting out prematurely or by stepping out of the frying pan of one heavy workload into the fire of another.

While this aspect gives Rose the ability to think on her feet and to act relatively quickly, the fixed Ascendant and fourth house fixed Mars suggest that Rose works best at her own pace and alone. While there is no burning desire to lead (and the Leo-Libra squares suggest perhaps some reluctance to lead for fear of not being liked), Rose certainly doesn't want to be anyone's flunky.

Mars opposes Rose's Midheaven, but again we're dealing with an out-of-sign aspect. There's a "quincunx-y" flavor to this opposition that gives an ability to compromise, but sometimes the compromise she makes is for the wrong reasons or to her own detriment. In Rose's case, I take this aspect to mean that she may have a bit of a blind spot when it comes to her home and those in it, and that she may put loyalty to family before personal fulfillment and loyalty to self at times. There would be an instinctive ability to anticipate the wants of those in her home and adjust to them. The trouble is, if she has to decide whether to fulfill her own drives or the wants of those around her, she may confuse their wants with hers and then wonder why her actions haven't taken her to a harmonious, happy place. (Note the Sun-Moon conjunction square Midheaven in this regard as well.) As a result of this confusion of "mine" and "thine," Rose may sometimes restrain herself (fixed Mars) from saying or doing things that really need to be said or done. Or she may choose an ineffective or inappropriate time to say or do them. Rose does admit to having had some problems in terms of airing dissatisfactions but has gotten better at this in recent years. Still, she has to fight the tendency to not make waves and to serve at her own expense.

As you might expect with Venus trine Mars in fire signs, Rose is an affectionate person who expresses a great deal of warmth. She likes men, and men find her attractive. Though Venus in Sagittarius can be flirtatious, the fixed Mars and the relatively strong Saturn in the eighth house make promiscuity unlikely. Venus in Sagittarius says Rose needs to believe in someone or something "higher" or "'nobler'" than her. The Venus-Neptune square reinforces the theme. Meanwhile, the seventh house position suggests a need for marriage or some sort of dependable partnership. Rose needs a committed relationship, and with that lovely trine she shouldn't have much trouble finding one. She needs to watch out for a tendency to idealize her mate, however. Otherwise, she could be badly hurt when he falls off his pedestal.

With Mars in Leo in the fourth, Rose would want to have some measure of control over her home environment. She'd also expect not to be taken for granted, and would want her home to be a reflection of who she is and what she's achieved in life—not only materially but in terms of inner security and spirituality.

What are her perceptions of what a man should be? He should be loving, reasonably successful, loyal, secure enough to be able to live with a strong woman, family-oriented and desirous of a nice home—at least according to her Mars.

How about sex? Well, with Mars in Leo trine Venus in Sagittarius, her approach is apt to be matter-of-fact and fairly direct. She'd enjoy the chase; with Venus square Neptune she'd also enjoy the candlelight and roses. She may also on occasion get swept up by the romance, trapped by her need to be needed or caught up in the pursuit of (or by) the wrong person. Venus in Sagittarius suggests Rose has a need for someone of different temperament and life experiences who can serve as a mirror or mediator. On the other hand, Mars in Leo in the fourth suggests that she wants the familiarity of a man whose upbringing and/or life experiences are similar to hers in at least some respects. In a woman's chart, Mars in the fourth is often indicative of marrying the boy next door or a childhood sweetheart. There don't appear to be any indications of sexual hang-ups.

In short, Rose should be able to attract men easily, and so long as she doesn't confuse love with dependency (Venus square Neptune, Mars in the fourth, Venus in the seventh and Sun, Moon, and South Node in the sixth suggest an awfully strong need to be needed!), she ought to be able to maintain relationships relatively well. Her wants (Mars) are pretty basic. They tally nicely with her needs (Venus). She both wants—and needs—male attention, compliments, and tangible proof that she is attractive to her mate. She wants to know that she pleases her mate because her own happiness in life is to a large extent dependent on her ability to please others—specifically her mate.

As stated earlier, Rose is an affectionate person who expresses a great deal of warmth. As you might expect from Venus trine Mars, she likes men, and men find her attractive. Though Venus in Sagittarius may flirt, the fixed nature of Mars and the fourth-seventh trine suggest that marriage is important to Rose and that she would be reluctant to live the single life by choice.

Rose married for the first time two days after her eighteenth birthday. Her husband was a professional musician, which tallies nicely with the Mars in Leo. Creative men—particularly actors, entertainers, and entrepreneurs or politicians—tend to be turn-ons for Mars in Leo women. He was nine years older than her. They had known one another for just under two years when they married. They were married for eighteen years, during which time they emigrated to Canada. They also had three children—one in the first few years of marriage, then two more in rapid succession nine years after the first.

What happened to this marriage? From what Rose says, it was not a bad marriage but rather one that was outgrown. At thirty-six, after the birth of the last of her three children in this marriage, she woke up one morning asking herself, "Is this all there is?" She decided to return to singing and was soon involved in performing in musicals. At this point her husband increasingly began to display jealous, possessive, and somewhat demanding behavior—not something a lady with Venus in Sagittarius and Mars in Leo is liable to want or need—especially hard on the heels of two pregnancies resulting in two children (one colicky, one an "easy baby") to look after.

In the course of her theatrical involvement, Rose and her husband became friendly with another couple—the other man we will call Joe. Rose and Joe went through an initial period of dislike (or perhaps attraction-repulsion), after which they became friends and eventually fell in love. Two divorces ensued, and Rose is now married to Joe. They have one child.

Joe is a Sun in Taurus. He is an executive with a large manufacturing company whose work involves traveling to many foreign countries as well as entertaining numerous foreign executives. Joe was born in the same country as Rose, has similar cultural interests, and shares Rose's desire for a physically comfortable lifestyle. His philosophical outlook is quite different from hers and he can at times be quite stubborn about his beliefs and wants. The marriage has had its ups and downs, but as Rose is monogamous and needs marriage, they are likely to remain together.

Looking at Rose's chart, it's clear that she has what it takes to be self-employed—a high energy level, ambition, and good common sense (as shown by the various Mars and Saturn placements and aspects). She has the potential to be a good businesswoman as well as quite a creative woman.

The quincunx-by-sign (albeit out-of-orb) Mars-Saturn combination suggests some difficulty in the career area due to work responsibilities conflicting with more personal wants, or perhaps familial responsibilities conflicting with career ambitions. Rose has climbed the ladder of success in real estate and the cosmetics industry, only to jump off the ladder. She has achieved a measure of success as a singer, though her name was never a household word. And she's climbed the ranks in volunteer organizations as well. Each time she has pulled back to pursue personal interests or attend to familial duties.

I have found that where there are quincunx vibrations, things rarely turn out to be what you thought they would. Additionally, with Mars in Leo there's a high level of ability to tolerate frustrations, but when the fuse blows, it blows—usually with a dignified retreat but also with an inner vow of, "Enough! I deserve better!" Remember, too, the earlier comments about Rose's tendency to step out of freedom-curtailing (according to Rose's perceptions) roles by undertaking other self-imposed responsibilities.

Of the career aptitudes suggested by Rose's Mars in Leo, she has dabbled with several. She has pursued the art of stained glass, acting, singing, has made and sold jewelry, and sold cosmetics. She has also sold real estate (fourth house Mars and Saturn in Capricorn) and taught gifted children and adults in various continuing education and after-school programs. From the time she was quite young, Rose wanted to be a singer. She began entertaining in clubs at a very early age and continued performing in her adult years in stage musicals and the like. However, she didn't really pursue a career in singing to the extent she could have. Why not?

While Rose has Mars in Leo—a good indicator for performing and entertaining, if not specifically singing—this Mars is in the fourth house. Additionally, Saturn (ruler of the tenth) is in the eighth. Both of these placements are relatively private and inner-directed. In other words, they clash with the limelight-seeking Leo.

From Rose's work history, there appears to be a pattern of almost but not quite reaching the top, of going far enough to prove that she can make it and then returning to the domestic arena only to re-emerge with another talent at another time. There is nothing inherently bad in this pattern if it works for Rose; she has the option of prioritizing her domestic and career potentials as she sees fit, and so long as she's comfortable, it would be a mistake for more career-oriented astrologers to berate her for not doing more. Likewise, it would not be constructive to discourage Rose's forays into the more public arena. She needs both. Presently, Rose is working part-time out of her home. This could be a way to reconcile her career and domestic needs.

Rose lives in a Mars-in-Cancer-city. This energy would to some extent motivate her to be aware of her security needs, her need to be needed, and her emotional wants. The city's Mars falls in Rose's third house. This might mean that Rose's need to get out of the house and involved in neighborhood/community activities would be a bit greater than otherwise, though the fact that

Mars is also in Cancer is going to bring to the fore that dichotomy between the need for recognition outside her family circle and her strong domestic/family drives. City Mars makes a sextile to Rose's Mercury, speeding up her thinking and communication responses and at least giving her an opportunity to be more assertive (and perhaps a bit quicker to debate pros and cons than she would otherwise be).

Rose has Jupiter in Cancer in the third house conjunct Pluto. This suggests a strong sense of curiosity as well as a very strong determination to make up her own mind. Rose's sources of opportunity would include neighbors, parents, and family members. She would appear to be lucky in terms of acquiring learning and security and to have good potential to succeed with creative hobbies or businesses operated from her home—particularly those involving metaphysics, sales, or desktop publishing.

Mars in Leo in the fourth makes Rose more motivated to think big and tackle big projects than to keep it small and simple. It also motivates her to do her own thing. However, the Leo placement says she needs attention and needs to mingle to make the most of her potential. She may at times be struck with inspirations that spring from deep within her, taking her by surprise but insisting on action on her part nonetheless. Again, home and community would appear to be the areas where she's most motivated to pursue opportunities, and pursuits connected with recreation, entertainment, or taking charge are liable to have the most appeal.

Mars is, by my standards, out-of-orb to a conjunction with Jupiter. Additionally, it is semi-sextile by sign. So timing can be just a wee bit off, or ideas just a bit too big to tackle single-handedly in many cases. She's more apt to be a bit too slow than a bit too quick on the uptake as Mars is fixed.

In the Koch system (which Rose finds works best for her) Mars is intercepted. Her sense of responsibility is therefore governed internally rather than externally, and to live comfortably with herself, Rose must be true to herself. The squaring Sun-Moon conjunction in Libra showing strong conditioning to please others certainly makes being true to herself a challenge at times. But she must be true to herself if she's to remain happy, healthy, and in control of her own identity.

An intercepted Mars in an inner-directed house suggests a great deal of potential for gaining insight into human motivation. This potential could be channeled quite nicely into theater, choreography, or other entertainment media.

Where a fourth house Mars often indicates people who leave home at an early age and burn the bridges behind them, intercepted Mars may leave home physically, but they never leave behind the early experiences that are a strong force for better or worse in their lives. It's not uncommon for people with a fourth house intercepted Mars to come into their adult years feeling "'hereditarily deprived." By this I mean that some factor in their upbringing has caused them to feel they are at a disadvantage. Perhaps there's an actual genetic physical problem to contend with. Perhaps they were poor and came up in life the hard way. Perhaps there was a "problem parent" whose reputation impinges on their own. In any case, these people feel they were born into a situation that put a strike against them, and they worry about this "strike" throughout their life.

When Mars is intercepted, the environment does not encourage security on some level. Ambitions are often challenged, motives and abilities questioned. You generally have to fight for what you get—and often fight with yourself to get it—and you're never entirely sure it's worth the fight. Ambition is not lacking, but goals are often so specialized that they're hard to fulfill. And in Rose's case, there's a suggestion that she feels pressure on herself to put making a home for her family ahead of expressing herself freely and fully. And indeed, there has been home/personal development conflict in Rose's life at many points.

Why is Rose here? Mars in Leo suggests that she wanted to take a more active role in the creative world this time around. To do this, she may need to learn how to get others to notice her talents. The talents may have been there previously, but she may have not been sure whether it would be "good" or "right" to push them (Libra Sun and Moon square Mars). Mars at 0° Leo suggests breaking new ground in the creative area. Perhaps in a previous life Rose was a financial benefactor to some theatrical company or talented individual or involved in the creative field in a very mundane way (as a stagehand, usher, etc.). Between that life and this one, my feeling is she has decided she is ready to use her own creativity rather than depend (Sun-Moon in Libra) on someone else to use it for her. The fourth house placement suggests a need to find a constructive outlet for her creativity as well as for her psychic/emotional energy. The Sun-Moon square to Mars suggests there may be some "yucky karma" to be dispersed in the various home environments—perhaps the residue of some sort of spiritual arrogance or "off-stage" prima donna behavior in a past life. Impatience may have been a past-life stumbling block, too, resulting in a need for an environment that demands she work for what she gets. Certainly there's been some sort of misdirected ambition in the past—not seriously misdirected, but perhaps wasted or underestimated.

The Sun-Mars square, which is out-of-sign, suggests Rose needs to get in touch with her power for doing rather than depending on someone else to do for her. The sextile-by-sign vibration suggests she will have an opportunity to do that.

The Moon-Mars square suggests there are self-control issues to be dealt with. Given the Leo-Libra vibration, there could have been problems stemming from excessive passivity, excessive ostentation, or passive-aggressive behavior. I would suspect there's been a fluctuation between over-control and under-control of temper, emotions, etc. Again, Rose has an opportunity to improve her balance in the area of self-control.

The Venus-Mars trine suggests that sensual/sexual/material issues have been reasonably well handled. There doesn't appear to be residual fear or a distaste for marriage, sex, or men in general. Nor does there appear to be a burning need to pursue these things, though the seventh house Venus suggests at least some need for marriage or partnership as a springboard for pursuit of her creative interests.

The Mars-Ascendant sextile suggests an opportunity to become more independent or more self-reliant in this lifetime. The Taurus-Leo vibration to this sextile says there may have been a bit too much complacency, dependence, or concern with what others might think in the past.

While this is by no means a full analysis of Rose, the person, it's a pretty thorough analysis of how she uses her energy, what turns her on, and what causes her to lose her patience. It also gives some clues to what might have motivated her to incarnate this time around and perhaps sheds some light on why she encountered certain types of people and circumstance at various points in her life.

The purpose of this exercise was not to prove that you need look no further than Mars for all the answers, but rather to show you how much information Mars will yield if you don't limit yourself to the old familiar "sex, selfishness, and temper" stereotypes. And this type of in-depth analysis can be done not just with Mars, but with each planet, to find repetitive themes and their minor variations.

While a blow-by-blow analysis of each planet in Rose's chart is beyond the scope of this book, I would urge you to go back over Rose's chart to add in the elements skipped over in this analysis. Then try this technique on your own chart or the chart of someone you know very well. While Mars is not the be-all and end-all of any chart, it's as good a place as any to start getting to know what makes a person tick —and why!

Afterword

In this book, I've tried to cover Mars from every conceivable angle, to go beyond the accurate but insufficient concepts of energy and motivation to show you how Mars affects personality, relationships, earning potential, and all sorts of intricate interplays between you and your environment. We've covered a lot of ground. I hope that in the process you've enjoyed yourself and learned something. But before we come to the end of this book, I'd like to leave you with one cautionary note. Having read this far, no doubt you'll go back to your charts with a new concept of the power of Mars, which is good; Mars is far too important to be overlooked. But at the same time, Mars isn't the only planet in the chart. What we've done in this book with Mars could just as easily have been done with the Moon, Jupiter, Uranus, or any other planet, for each planet (as well as the angles and Nodes) has an intricate function. No one planet "makes" us what we are; it's the interrelationship between the planets, signs, houses, and aspects that gives us our unique characteristics of personality. And these in turn are modified by our free will.

The sum of the information given in a chart is greater than what can be indicated by an individual planet, sign, house, or whatever. This is where synthesis comes in—and this is where all astrology books, no matter how good, fall down. For to write about any facet of the chart, you must take it out of context, simply because the possibilities in astrology are so endless that there's frankly no one all-purpose interpretation for anything! This is why in every book you read, no matter how much of the material "fits" your chart, you'll always run into "clinkers," things that have absolutely no relevance to your feelings and behavior-even though you have the aspect present in your chart. When you run into these clinkers, don't just assume they must fit because the author says so; on the other hand, don't assume the author is wrong. Instead, go back to your chart and find out *why* the information doesn't fit. What makes you different? What aspects or sign/house placements could offset or alter the behavior potentials described? Maybe you're not being objective about yourself. Ask a friend to read the interpretation you feel doesn't fit and to tell you honestly if he or she feels that what has been written is applicable to you. Maybe you don't see yourself clearly in this area. Or maybe the author is just plain wrong! We all have our blind spots.

But books—even the best of them—can take you only so far. The rest of your knowledge must come from actually working with charts, from learning first-hand that Mars in Pisces in the first house isn't necessarily energetic even though Mars in the first house is supposed to be physically active. When you learn to see the whole rather than each small bit of information, then you're an astrologer, and only then will you be able to grasp the full meaning of the chart.

Appendix I

Mars Affinities

Many things and qualities are ruled by more than one planet. Those mentioned here are strongly tied to Mars, but may have links to other planets as well. For further information, consult a good keyword listing such as *The Rulership Book* by Rex E. Bills.

Sign: Aries, which is a masculine, cardinal, fire sign

House: First

Sports: Hunting, shooting, riding, boxing, football, hockey, wrestling, car rallying

Flowers: Red roses, red gladioli, geraniums, thistles, honeysuckle

Scents: Cedar, pine, cypress, attar of roses, acidic scents, sharply astringent scents

Day of the Week: Tuesday

Metal: Iron

Gems: Bloodstone, diamond

Colors: Scarlet, cherry, magenta, wine, red

Number: Three

Careers: Dentistry, barbering, electronics, cashiering, stenography, butchering, undertaking, psychology, psychiatry, exploring

Period of the Year: March 21 to April 19

Herbs: Aloes, capers, cinnamon, coriander, hops, arnica, wormwood, allheal

Plants/Trees: Cactus, ironwood, hemlock, pepperwort, pine, witch-hazel, and all thorny trees

Wines: Rhine wines, champagne

Typical Aries Personalities: Joan Crawford, Arturo Toscanini, Charles Wilkes (explorer), Otto von Bismarck, Thomas Jefferson

Character Traits: Ambition, daring, dissent, initiative, larceny, temerity, enterprise, courage, selfishness, lack of subtlety, impulsiveness, assertiveness, leadership, mental or physical domination, dynamism, pride

Miscellany: Abrasions, blisters, clamor, enemies and dislikes, fireplaces, hoes, massacres, noise, pornography, rapes, sandpaper, scissors and shears, squabbles and fistfights, wrenches, daggers and knives

Appendix II

Sun-Mars Quickie Synastries

This is a handy-dandy guideline for determining what might occur if you choose to see that delightful stranger again. As you can see, the descriptions given don't have an awful lot of depth. They're not meant to since, as you probably know, you can't do very much on the basis of Sun sign alone. My main reason for including this is that many times clients (and occasionally beginning students) ask for quickie guidelines of this nature, and these seem to work better than the Sun sign comparisons you find in the magazines.

To use these tables, you need to know two things: your Mars sign and the other person's Sun sign. Look for your Mars sign. When you've found it, find the other person's Sun sign. Read what's there. That's it.

It's interesting how, after only two or three meetings—or even, occasionally, after one meeting—you can see the synastry operating along these lines. But again, this is just one facet of a relationship. Before you throw in the towel (or move in together, as the case may be), you'll want to look at other facets of the relationship. Use these guidelines as a starting point when you have nothing else to go on.

To give you some idea of how this would work, we'll consider an example: A woman with Mars in Libra meets a man who has Sun in Virgo. According to our guidelines, much would depend on the overall compatibility, so if she's interested, she could pursue the relationship and, when she got to know him better, get his complete data. As our guidelines suggest that she'd be the dominant partner, she'd be wise to give him some overt encouragement if she's interested. With Mars in Libra, she'd be capable of doing this.

Okay... now, suppose our subject in question is a man with Mars in Gemini. Combine this with a woman's Leo Sun and we have a potentially friction-producing combination. He'd expect and allow her to make the first move, but eventually he might get fed up with her being the dominant partner. Indeed, whether or not this relationship would get beyond three or four dates would depend very much on this overall compatibility. So from this brief analysis, regardless of which party did the comparing, things look "iffy" and a lot will depend on their initial attraction to one another in terms of whether or not they'll give the relationship a chance to develop.

As you can see, this technique is a sketchy sort of guideline and leaves a lot of question marks at best. But it does have two advantages. First of all, it gives you some idea of whether or not it's worthwhile to see the other person again. (In "iffy" cases, as with our lady friend just described, I usually advise a second meeting if only to clarify your feelings.) Second, this technique gives you some idea of just how much initiative to take. For example, in the case just described, it's clear that the lady is going to have to take the initiative if she wants things to go anywhere, as without knowing the man's Mars sign she appears to be the dominant partner. In other cases, where the other person is dominant, this tells you that you'll just have to bide your time and wait for some sign of interest or encouragement from the other person.

Mars in Aries

Aries Sun: There's potential for a successful relationship, but this combination will undoubtedly generate a lot of energy so you'd better find a constructive way of expending it—quick! Expect some stormy periods in your relationship, but if you're motivated to make it work, it will.

Taurus Sun: Potentially good, but not necessarily lasting. Although you're generally willing to forget hassles and start afresh, your friend may not be!

Gemini Sun: You're probably the dominant partner, and the other person will generally be content to follow your lead. You'll enjoy his or her witty conversation, if nothing else.

Cancer Sun: The other person will try to dominate, with friction probably resulting. He or she expects loyalty. Are you willing to give it?

Leo Sun: This can be a mutually satisfying relationship if other factors in the chart agree. Certainly there is no lack of passion. Whether it eventually ends or not will depend on what else you've got going for you.

Virgo Sun: You're be the dominant partner. At first the other person won't mind and the relationship can be fairly enjoyable. But eventually you could begin to take him or her for granted and treat the person more like a servant than a lover. At this point, the other person would become resentful and there would be problems.

Libra Sun: Extremely pleasant or extremely difficult, with not much in between. You're the dominant partner, but the other person would still probably gain more from the relationship than you do. He or she could easily become quite dependent on you, which might cause you problems in the long run.

Scorpio Sun: This is a fight-producing combination that fluctuates from terrific to terrible.

Sagittarius Sun: You're the dominant partner. Okay as a friendship, but not particularly easy romantically. Some quarrels are inevitable and Sagittarius is apt to brood when he or she doesn't win—which is often the case with this combination.

Capricorn Sun: If you're both persistent and motivated to make the relationship work, it can, but even at its best there is some friction.

Aquarius Sun: You're the dominant partner sexually; the other person dominates intellectually. There's potential here for a good, complementary relationship.

Pisces Sun: You're the dominant partner, but the other person's mellowness could soften you up in a beneficial way. You benefit from his or her sympathetic streak and empathy, and he or she benefits from your more action-oriented approach to life.

Mars in Taurus

Aries Sun: There's potential for a relationship between equals, though Aries may find you a tad more phlegmatic than he or she likes.

Taurus Sun: There's extremely good potential in that you probably have mutual likes and dislikes. But after the initial excitement of being in love fades, and the rosy glasses have been put away, you may be less than delighted to see that you also have a stubborn streak in common.

Gemini Sun: You'll be the dominant partner initially, but it would be Gemini who decides when it's time to get up and move on. This is apt to be either a short-lived relationship or else not entirely satisfactory in some way.

Cancer Sun: The other person would soften up your inflexible streak, and you'll provide stability for him or her. Potentially good.

Leo Sun: The Leo builds your confidence. You're inclined to dominate, all the while letting Leo think he or she is actually boss. Probably both of you are stubborn, so some friction is possible.

Virgo Sun: You'll dominate, with the other person generally following willingly. This is often what I call a "learning experience" relationship; eventually one or both of you may outgrow it, but it could be quite good while it lasts.

Libra Sun: This relationship often begins with a strong physical attraction—and sometimes there's not much else! If the Libra truly respects you, it *could* work, but it's not apt to be an entirely easy relationship.

Scorpio Sun: You'll make the other person "prove" himself or herself before getting too involved. Scorpio dominates over the long term, something you're apt to resent and against which you may eventually rebel.

Sagittarius Sun: You'll dominate, probably often giving the other person a hard time and expecting him or her to do a lot of the changing in the relationship. Not good.

Capricorn Sun: The other person dominates, but normally he or she will be eager to please, so this can work pretty well.

Aquarius Sun: You'll dominate, leading to friction. Strike one in terms of a lasting relationship.

Pisces Sun: The other person will soften you up. You'll give him or her the reliability he or she needs. Good.

Mars in Gemini

Aries Sun: The other person will dominate with your cooperation. Good relationship potential; you can be extremely loyal to one another, yet you allow one another a measure of freedom.

Taurus Sun: The other person will dominate, with some friction resulting. Whether or not the relationship will work in the long run depends on other factors in the comparison. Should the relationship run into difficulties, there's often an unwillingness or inability to change, and outside help may be needed to regain your objectivity.

Gemini Sun: You can be enchanted with one another and are apt to be on the same wavelength sexually. So this can be favorable if other factors in the comparison are in harmony.

Cancer Sun: The other person will dominate with some friction resulting. The main danger is that Cancer doesn't enjoy experimenting and you do. Cancer may push for a permanent relationship before you feel ready to give up your freedom, which can cause you to feel threatened. Again, other factors in the comparison must be considered.

Leo Sun: The other person will dominate with your cooperation. Decent romantic potential. Seems to work best if the Leo is older or more experienced than you.

Virgo Sun: The other person dominates. Friction results. A difficult relationship to maintain.

Libra Sun: The other person dominates with your cooperation. Good potential if other factors agree.

Scorpio Sun: The other person dominates. Friction results. This often smacks of "can't live with 'em; can't live without 'em."

Sagittarius Sun: A lot depends on age in this case. This could work if you've both done your experimenting and are ready to try something more stable and steady. Otherwise it tends to be a potentially difficult combination involving much friction.

Capricorn Sun: The other person dominates. Friction is apt to result. Somebody generally winds up feeling bad regardless of whether the relationship lasts.

Aquarius Sun: The other person dominates with your cooperation. Could be quite nice.

Pisces Sun: The other person dominates, leading to friction. Probably a frustrating and short-lived relationship.

Mars in Cancer

Aries Sun: You dominate, leading to friction. Extremely romantic at times, but can also be extremely turbulent.

Taurus Sun: The other person allows you to dominate. Good potential as long as you're in agreement about kids.

Gemini Sun: You dominate, leading to some friction. Gemini talks about ex-loves and you get all defensive and contentious. Much depends on other aspects in the comparison.

Cancer Sun: Very good potential, assuming the rest of the comparison agrees.

Leo Sun: You're going to be determined to get your own way. If you dominate in an affectionate, low-key way that doesn't threaten the other person's ego, this can work.

Virgo Sun: You dominate, especially if Virgo is younger or less experienced than you. Generally good relationship potential.

Libra Sun: You dominate. Friction results. A hard relationship to maintain.

Scorpio Sun: You dominate. Scorpio appreciates your sensitivity but may not always understand you. This tends to be a sporadic relationship, not so much because of friction as because of a lack of shared interests or goals. Probably short-lived.

Sagittarius Sun: You dominate. Friction results. This tends to be on-again, off-again.

Capricorn Sun: You dominate, especially if the other person is younger than you. Friction results, though if you're both mature you can deal with it successfully and wind up with a reasonably harmonious relationship.

Aquarius Sun: Whoever is more intelligent will dominate. If your intelligence levels are equal, you'll dominate. Either way, friction results and you can wind up with a "can't live with 'em, can't live without 'em" relationship.

Pisces Sun: You dominate, but Pisces probably won't mind. You both want marriage and are motivated to make your relationship last. Fairly good potential.

Mars in Leo

Aries Sun: You'll want to go out with this person several times before you make up your mind, just to be sure there's something more than physical attraction (which tends to be plentiful). This tends to be a relationship between equals. Good potential.

Taurus Sun: Probably a slow-moving relationship with a lot of initial caution on both sides. Taurus will want to be in charge of the money and may criticize your spending habits. Care would have to be taken to keep interference with one another's individuality and preferred routine to a minimum. Taurus brings out your tenderness, but at the same time there is some danger of you interfering with his or her self-expression. This seems to work best if the woman has the Sun in Taurus and the man has the Mars in Leo.

Gemini Sun: You dominate. In fact, Gemini is often initially attracted to your strength and determination. Good potential.

Cancer Sun: The other person dominates using his or her charm and powers of self-preservation to gain the advantage. You both have a very strong need for harmony in your love life, which can work in your favor. Much depends on the overall compatibility.

Leo Sun: Not so good. You both like to show off and can compete in a counterproductive way.

Virgo Sun: Virgo needs his or her love to be up on a pedestal. You rarely mind this. When the honeymoon's over, the relationship can become more equal and based on mutual needs, but if there's nothing much holding the synastry together, it could just as easily be outgrown.

Libra Sun: Good potential for a thriving relationship between equals unless the comparison shows strong contraindications.

Scorpio Sun: The other person dominates. Ranges from fairly good to terrible, depending on the overall compatibility. Ask yourself if you can relax with one another out of bed as well as within.

Sagittarius Sun: You're generally attracted, but the attraction may or may not last. You dominate. There can be sexual difficulties if either chart shows hang-ups in that area.

Capricorn Sun: You can talk yourself in and out of this relationship several times due to the fact that it may be beneficial in some respects (particularly materially or in terms of status) and unfulfilling in others (like the less tangible aspects). Capricorn would dominate. Much depends on the overall compatibility; this could, in itself, go either way.

Aquarius Sun: You dominate; friction results. You're both probably extroverts, which is a plus, but a lot depends on whether you can indulge one another without feeling you're losing your dignity or independence.

Pisces Sun: A relationship between equals, although when it's a Sun in Pisces woman with a Mars in Leo man, the woman will have little difficulty wrapping the man around her little finger if she decides to do so. I find this situation slightly better than Sun in Pisces man with Mars in Leo woman, though, as in this case the woman can at times make the man feel henpecked.

Mars in Virgo

Aries Sun: The other person dominates. Much depends on the overall compatibility. Aries would have to be very understanding of your defense mechanisms as well as your goals.

Taurus Sun: Neither of you is inclined to rush into things, so this has got to be a slow-developing relationship. Taurus generally dominates with your cooperation. Good potential.

Gemini Sun: You dominate; friction results. You're both good at disguising your feelings with a lot of extraneous verbiage, which makes real closeness and understanding difficult.

Cancer Sun: Attraction is common. Cancer dominates. Good potential unless the overall comparison shows otherwise.

Leo Sun: A relationship between equals, assuming your intelligence levels are roughly the same. Some friction, but generally not of major proportions. Worth trying.

Virgo Sun: Very good potential. Worth trying, and worth getting the other person's complete data.

Libra Sun: The other person dominates. Much depends on the overall compatibility.

Scorpio Sun: The other person dominates. Decent romantic potential, though often you meet through work or get involved in some sort of work-related partnership.

Sagittarius Sun: You dominate; friction results. Not an easy relationship to maintain.

Capricorn Sun: Not apt to be a good romantic relationship in the long run, although it can be advantageous to both of you in the short term.

Aquarius Sun: A relationship between equals if it lasts. Generally this one is very shaky at the outset and even when it's lasting there's usually at least one break-up between the introduction and the altar. You'd best do a compatibility if you make it beyond the second or third date.

Pisces Sun: The other person dominates; friction results. Probably far from the ideal relationship, but for some reason it's often a lasting one.

Mars in Libra

Aries Sun: The other person dominates. Friction results. This is usually evident from the outset. You enjoy the courtship period and want soft music, candlelight and roses; Aries will humor you initially, but wants to get down to basics—like having his or her needs met—a lot quicker than you may like.

Taurus Sun: This can be a relationship between equals based on mutual needs so long as you're sensitive to one another's vulnerabilities.

Gemini Sun: You dominate. If you respect each other's ideas and have similar intelligence levels, potential is good. Either one of you can easily attract someone else any time you like, though, so if boredom or difficulty sets in, you may not be highly motivated to work things through. Some Saturn contacts in the comparison would compensate for this.

Cancer Sun: The other person dominates; friction results. A hard relationship to maintain.

Leo Sun: The other person dominates and takes the lead sexually with your approval. Good potential; seems to be especially good when it's a Sun in Leo man with a Mars in Libra woman.

Virgo Sun: You dominate. Much depends on the overall compatibility. Not stress-free, but not necessarily bad.

Libra Sun: Good potential assuming you have mutual interests and other links in the comparison.

Scorpio Sun: A relationship between equals. Not entirely stress-free, but it's worth a try.

Sagittarius Sun: You dominate, but you're generally helpful to Sagittarius and vice versa. Good potential.

Capricorn Sun: The other person dominates. Friction results. Not an easy relationship.

Aquarius Sun: You dominate. Good potential, especially if you're both over thirty.

Pisces Sun: You dominate. Much depends on the overall compatibility. As a rule, this one is "on-again, off-again."

Mars in Scorpio

Aries Sun: This can be a relationship between equals, but it's one you should be careful with as it's apt to be highly volatile at times. Not recommended if you believe love is never having to raise your voice in anger!

Taurus Sun: You dominate. Friction results, but on the other hand your energy levels and passion levels are similar and this may help hold things together.

Gemini Sun: You dominate. Much depends on the overall compatibility. You can be very jealous, and chances are good that Gemini won't tolerate this very well or very long.

Cancer Sun: The other person dominates. Apt to be an "on-and-off" type of relationship. You'll have to learn to be tolerant of one another's little quirks and understanding of one another's pet peeves.

Leo Sun: You dominate. Friction results. Leo won't display an awful lot of tolerance about your brooding, and you're both inclined to turn resentful if your egos are thwarted.

Virgo Sun: You dominate. You can hurt one another's feelings at times, so a great deal depends on how you handle your anger and whether or not either of you is inclined to hold a grudge. If you're both mature, however, this can be a very happy match.

Libra Sun: A relationship between equals. Some friction, but unless the compatibility shows otherwise, you can probably handle it.

Scorpio Sun: Can be a good combination; check to see if other aspects concur.

Sagittarius Sun: You dominate. Much depends on the overall compatibility. Expect at least a little friction.

Capricorn Sun: Potential for a relationship between equals in terms of power. For some reason, this combination tends to produce a very private "we want to be alone" type of relationship.

Aquarius Sun: You dominate. Friction results. Not an easy relationship to maintain.

Pisces Sun: Potential for good sharing and a torrid sex life. Worth a try.

Mars in Sagittarius

Aries Sun: The other person dominates. Good potential, with enough differences in interests, et cetera, to keep things interesting.

Taurus Sun: The other person dominates. Much depends on the overall compatibility. Taurus is apt to find you kind of reckless; you may find Taurus a trifle stodgy.

Gemini Sun: You dominate. Friction results, but if you're both bright, independent people, this could work.

Cancer Sun: The other person dominates. Much depends on the overall compatibility. This one tends to be "on-again, off-again."

Leo Sun: The other person dominates. Can lead to friction, but not necessarily enough to seriously impair compatibility. This one works best if there's a shared hobby or business interest.

Virgo Sun: The other person dominates. Friction results. Not an easy relationship to maintain.

Libra Sun: The other person dominates and often seems to spend a lot of time "testing" your worthiness. If there are any inter-aspects to keep you together, you'll probably remain faithful, but at times it won't be easy. This seems to work best when it's a Sun in Libra woman with a Mars in Sagittarius man.

Scorpio Sun: You dominate. Much depends on the overall compatibility. Expect at least a little friction.

Sagittarius Sun: Good potential assuming there are other contacts in the compatibility.

Capricorn Sun: The other person dominates. Much depends on the overall compatibility.

Aquarius Sun: The other person dominates. Good potential and worth a try.

Pisces Sun: The other person dominates. Friction results. Not awfully easy.

Mars in Capricorn

Aries Sun: Can be a relationship between equals, but it's not always easy. You're apt to be introverted; Aries is apt to be extroverted. Your dark depressions or bad moods are apt to get on Aries' nerves.

Taurus Sun: The other person dominates. Good potential for a lasting relationship.

Gemini Sun: Difficult, especially if it involves a Sun in Gemini male and a Mars in Capricorn female. In any case, your timing tends to be out-of-sync and friction results. You dominate.

Cancer Sun: Cancer dominates and will delight in making a home for you and taking care of you, which is fine if that's what you want. If it isn't, friction results.

Leo Sun: You dominate. Much depends on the overall compatibility. Tends to be "on-again, off-again.

Virgo Sun: You dominate. Good potential and certainly worth a try!

Libra Sun: You dominate. Friction results. This is a hard relationship to maintain.

Scorpio Sun: You dominate. Good potential, especially if you're both over thirty.

Sagittarius Sun: You dominate. Much depends on the overall compatibility. Expect at least a little friction.

Capricorn Sun: Generally a beneficial relationship for as long as it lasts. And it can last.

Aquarius Sun: You dominate. Much depends on the overall compatibility. This one can go either way.

Pisces Sun: You dominate. Good potential for a satisfactory relationship.

Mars in Aquarius

Aries Sun: The other person dominates with your cooperation. You both demand a lot of freedom in relationships, so you allow one another breathing space. Good potential.

Taurus Sun: You both prefer to be in control. Chances are Taurus will dominate, with friction resulting. Not easy.

Gemini Sun: You dominate. Your desires for physical comforts and mental stimulation are similar. Good potential.

Cancer Sun: You can take an interest in one another's well-being, but chances are you'll be a bit more involved in outside interests than Cancer likes. Cancer will try to dominate. Much depends on the overall compatibility.

Leo Sun: The courtship phase tends to be romantic and great fun. Eventually, Leo dominates. Friction results. This one can go either way.

Virgo Sun: Can be a relationship between equals based on mutual needs which override more superficial wants. Ideally there would be a shared interest of some sort.

Libra Sun: The other person dominates. Not so good if it's a Mars in Aquarius female and a Sun in Libra male. Otherwise, decent potential.

Scorpio Sun: The other person dominates; friction results. Not an easy relationship to maintain.

Sagittarius Sun: Can be a relationship between equals. Good potential and well worth exploring.

Capricorn Sun: The other person dominates. Much depends on the overall compatibility.

Aquarius Sun: Good potential if there are also mutual interests. Worth exploring further.

Pisces Sun: Good potential if there are also mutual interests. Otherwise there could be friction.

Mars in Pisces

Aries Sun: The other person dominates. Much depends on the overall compatibility. This combination seems to work best when there's a shared business or recreational interest playing a prominent role in the relationship.

Taurus Sun: Taurus dominates with your cooperation and has just the right amount of the iron hand within the velvet glove. Good potential for a lasting relationship.

Gemini Sun: Gemini loves freedom; you love togetherness. You dominate. Friction generally results and Gemini generally leaves.

Cancer Sun: The other person dominates. Good potential, though both of you have tendencies to make mountains out of molehills and this could at times cause problems.

Leo Sun: Probably you'll recognize each other's needs; whether or not you can fulfill them depends on the overall compatibility. You prefer being possessed to being possessive; Leo prefers the reverse. So in this sense at least there's potential.

Virgo Sun: You dominate. You tend to be attracted to one another, but this pairing is seldom easy.

Libra Sun: The other person dominates. Much depends on the overall compatibility. This one tends to be "on-again, off-again."

Scorpio Sun: The other person dominates. This can have good romantic potential and is worth exploring.

Sagittarius Sun: You'll recognize each other's needs. With work, you may be able to fulfill them.

Capricorn Sun: The other person dominates. This seems to work best when it's a Sun in Capricorn male and a Mars in Pisces woman; when it's the other way around, there are often difficulties stemming from different attitudes toward finances.

Aquarius Sun: If you have equal intelligence levels, there's decent potential. Otherwise, there can be friction.

Pisces Sun: Quite good potential. Worth exploring further. You'll probably enjoy looking after one another.

Appendix III

Mars Positions for Various Cities

The following material comes from either *Horoscopes of the USA and Canada* by Marc Penfield (Tempe, AZ: AFA, 2005); or Horoscopes of U.S. States and Cities by Carolyn R. Dodson (San Diego, CA: ACS Publications, 1975). In Chapter 4, I listed Mars positions for various cities according to my own chart preferences. In this section, I've included Mars positions from both books so you can compare.

In the following table, I have indicated which data is from Penfield and which is from Dodson. Cities are listed in alphabetical order.

Table 2. Dodson/Penfield Mars Positions

City	Dodson	Penfield
Akron, OH	02 Pisces 26	08 Libra 19
Alexandria, VA	00 Capricorn 19	28 Sagittarius 27R
Anchorage, AK	27 Capricorn 13	19 Leo 29
Asheville, NC	14 Scorpio 42	25 Aquarius 56
Aspen, CO	not listed	17 Aries 55
Atlantic City, NJ	06 Virgo 03R	06 Virgo 01R
Augusta, ME	24 Aries 38	12 Capricorn 17
Austin, TX	04 Aquarius 54	28 Virgo 45

City	Dodson	Penfield
Baltimore, MD	23 Cancer 20	17 Cancer 27
Birmingham, AL	2 Aquarius 33	07 Libra 07
Brandon, MAN, CAN	not listed	11 Leo 50
Brigham City, UT	11 Cancer 34R	not listed
Buffalo, NY	27 Aquarius 45	10 Libra 09
Calgary, ALTA, CAN	not listed	25 Sagittarius 16
Charleston, SC	13 Aries 05	01 Sagittarius 11
Charlotte, NC	29 Aries 38R	29 Aries 46R
Charlottetown, PEI, CAN	not listed	07 Leo 59
Chattanooga, TN	29 Leo 25	13 Leo 16
Chicago, IL	15 Virgo 38	27 Virgo 58
Cincinnati, OH	19 Sagittarius 04	07 Capricorn 34
Cleveland, OH	22 Scorpio 45	17 Sagittarius 15
Columbia, SC	09 Capricorn 15	24 Gemini 43
Columbus, OH	17 Taurus 27	12 Aries 40
Dallas, TX	19 Libra 06	29 Libra 52
Dawson, YUK, CAN	not listed	01 Gemini 00
Daytona Beach, FL	not listed	20 Virgo 00R
Denver, CO	20 Libra 57	07 Aquarius 04
Des Moines, IA	09 Cancer 31	19 Sagittarius 46R
Dodge City, KS	not listed	27 Cancer 54
Dover, DL	05 Gemini 57	17 Scorpio 16
Edmonton, ALTA, CAN	not listed	01 Libra 51
Fairmont, WV	28 Scorpio 29	not listed
Fall River, MA	27 Gemini 43	28 Leo 05
Fargo, ND	14 Virgo 28	06 Libra 53
Fort Wayne, IN	19 Pisces 47	05 Capricorn 19
Frankfort, KY	27 Pisces 54	16 Virgo 04R
Fredericton, NB	not listed	04 Pisces 52
Grand Rapids, MI	08 Cancer 04	27 Capricorn 44
Halifax, NS	not listed	03 Capricorn 04R
Hamilton, ONT, CAN	not listed	16 Virgo 30
Harrisburg, PA	05 Scorpio 54	27 Aries 21
Hartford, CT	16 Cancer 34R	02 Gemini 31

City	Dodson	Penfield
Hastings, NE	21 Taurus 43	not listed
Honolulu, HI	12 Capricorn 13	28 Capricorn 11
Houston, TX	02 Virgo 32	03 Cancer 07
Huntington, WV	05 Libra 57R	06 Libra 03R
Jacksonville, FL	05 Capricorn 26	13 Virgo 34
Jamestown, VA	not listed	01 Cancer 40
Juneau, AK	01 Gemini 45	17 Virgo 23
Kansas City, MO	09 Leo 50	23 Gemini 08
Las Vegas, NY	29 Pisces 15	15 Scorpio 02R
Lewiston, ID	25 Cancer 40R	18 Gemini 46
Little Rock, AR	29 Libra 54	23 Scorpio 02
London, ONT, CAN	not listed	02 Aquarius 20
Long Beach, CA	15 Sagittarius 40	15 Sagittarius 06
Los Angeles, CA	17 Capricorn 48	17 Capricorn 48
Louisville, KY	03 Sagittarius 32	08 Gemini l9
Mankato, MN	15 Pisces 43	not listed
Miami, FL	18 Taurus 24	08 Pisces 05
Missoula, MT	26 Aquarius 24	13 Gemini 22R
Mobile, AL	13 Scorpio 03	18 Scorpio 47
Monterey, CA	not listed	11 Aries 25
Montgomery, AL	05 Leo 38	05 Leo 37
Montpelier, VT	14 Capricorn 41	26 Virgo 30
Montreal, QUE, CAN	not listed	10 Pisces 14
Nashville, TN	06 Leo 43	29 Taurus 34
Natchez, MS	not listed	24 Taurus 59
New Bedford, MA	05 Aquarius 26	24 Aquarius 53
New Haven, CT	10 Libra 32	21 Sagittarius 42
New Orleans, LA	02 Leo 26R	15 Aries 50
Newport, RI	2O Cancer 58	14 Gemini 45
New York, NY	29 Sagittarius 17	08 Cancer 24
North Platte, NE	16 Pisces 18	27 Cancer 15
Oakland, CA	11 Leo 04	00 Leo 44
Oklahoma City, OK	28 Scorpio 49	17 Taurus 50
Oshkosh, WI	22 Pisces 29	02 Leo 29

City	Dodson	Penfield
Ottawa, ONT, CAN	not listed	10 Virgo 00
Philadelphia, PA	05 Libra 34	14 Virgo 20
Phoenix, AZ	27 Capricorn 13	25 Leo 14
Pittsburgh, PA	04 Scorpio 46R	23 Sagittarius 55
Pittsfield, MA	03 Taurus 49	03 Taurus 44
Pocatello, ID	22 Taurus 54	22 Taurus 30
Portland, OR	15 Pisces 13	06 Aquarius 31
Portsmouth, NH	06 Taurus 16	08 Capricorn 37
Price, UT	24 Capricorn 42	not listed
Quebec City, QUE, CAN	not listed	16 Aquarius 09
Raleigh, NC	24 Pisces 54	19 Virgo 02R
Rapid City, SD	05 Scorpio 54	28 Aries 27
Regina, SASK, CAN	not listed	24 Cancer 32
Reno, NE	24 Gemini 12	21 Aries 25
Rochester, MN	26 Scorpio 33	01 Libra 55
Rock Springs, WY	15 Capricorn 59	10 Gemini 38
Sacramento, CA	24 Gemini 14	26 Libra 47
Saint Augustine, FL	20 Gemini 05	20 Gemini 08
Saint John, NB	not listed	23 Aquarius 30
Saint John's, NFLD, CAN	not listed	21 Capricorn 22R
Salem, OR	19 Cancer 07	09 Gemini 24
San Antonio, TX	12 Aries 57	H Gemini 03
San Francisco, CA	14 Cancer 22	16 Gemini 34
Sante Fe, NM	22 Leo 59	12 Sagittarius 38
Saskatoon, SASK, CAN	not listed	26 Gemini 40
Savannah, GA	01 Virgo 29	10 Aries 55
Sioux Falls, SD	22 Aquarius 25	10 Leo 08
Spokane, WA	14 Cancer 30R	08 Taurus 41
Springfield, IL	10 Aquarius 24	01 Aries 06
Sudbury, ONT, CAN	not listed	23 Taurus 54
Syracuse, NY	04 Taurus 16	12 Leo 31
Toledo, OH	26 Leo 13R	16 Gemini 28
Toronto, ONT, CAN	not listed	16 Cancer 33
Topeka, KS	24 Pisces 10	11 Capricorn 40

City	Dodson	Penfield
Trenton, NJ	04 Capricorn 33	02 Sagittarius 04
Tucson, AZ	15 Sagittarius 46	24 Libra 20
Tulsa, OK	01 Libra 34	26 Cancer 29
Twin Falls, ID	24 Scorpio 14R	not listed
Victoria, BC	not listed	09 Sagittarius 36
Wheeling, WV	01 Aquarius 02	00 Scorpio 39
Whitehorse, WK, CAN	not listed	23 Gemini 06
Wichita, KS	23 Virgo 05R	02 Aries 05
Wilmington, DE	01 Sagittarius 30	18 Sagittarius 22

As you will see, for some cities the Mars positions vary by only a minute. In others they vary by several signs. How can this be? The difference is in the authors' philosophy of what constitutes a birth. Dodson has, in almost all instances, used the data for the legal incorporation of the city and used noon, local mean time where specific times were unavailable. Penfield has used either the dates when permanent settlement began or city founding dates, and has made some attempt at rectification in many instances.

So which chart is right? In my experience, both charts are right, and there are probably other equally "right" charts for many of the cities listed. I think it's common to have a preference for one chart over another, but I also think this is more of a philosophical issue than an academic one. For example, I tend to see the appearance of a permanent settlement as a sort of birth; the incorporation, or formal acknowledging is to me sort of like the filing of a birth certificate or a baptism or something. Other people see the settlement as conception and the incorporation as birth. In any case, I find both charts useable, but in different ways. Penfield's charts tend to give me a psychological feel for the city and its vibes, while Dodson's charts provide a more concrete and mundane perspective relating to the economy, politics, school systems, and so on. Both types of information are valuable, but at different times and for different reasons. People planning a vacation, for example, tend to be mainly concerned with the vibes and not at all interested in the mundane. People relocating, on the other hand, need to consider both—and if employment is an issue, the Dodson chart should probably be emphasized.

Those who want to get really detailed information can also look at state charts, even comparing state to city charts. Both of the above-mentioned references include state charts, which also vary in some instances.

www.ingramcontent.com/pod-product-compliance
Lightning Source LLC
Chambersburg PA
CBHW082316230426
43666CB00036B/2754